John Edward Gray

Synopsis of the Species of Starfish in the British Museum

John Edward Gray

Synopsis of the Species of Starfish in the British Museum

ISBN/EAN: 9783337410025

Printed in Europe, USA, Canada, Australia, Japan

Cover: Foto ©berggeist007 / pixelio.de

More available books at **www.hansebooks.com**

SYNOPSIS

OF THE

SPECIES OF STARFISH

IN THE

BRITISH MUSEUM.

(WITH FIGURES OF SOME OF THE NEW SPECIES.)

BY

JOHN EDWARD GRAY, Ph.D., F.R.S., V.P.Z.S., F.L.S., etc.

LONDON:

JOHN VAN VOORST, 1 PATERNOSTER ROW.

MDCCCLXVI.

PRINTED BY TAYLOR AND FRANCIS,
RED LION COURT, FLEET STREET.

INTRODUCTION.

In the 'Annals and Magazine of Natural History' for November and December 1840 (vol. vi. pp. 165 & 275) I published a " Synopsis of the Genera and Species of the Class Hypostoma (*Asterias*, Linn.) ; " and in the ' Proceedings of the Zoological Society ' for May 25, 1847 (see part xv., and Ann. & Mag. N. II. xx. p. 193), I described some new genera and species of *Asteriadæ*.

The former paper was prepared as a prodromus to a work which I intend to prepare, containing figures of all the genera and of most of the species of this class.

The Plates now published were then prepared; but I deferred the continuation of the work at the urgent request of Professor Agassiz, who wished me to unite with him, and bring out the work under our joint names, forming part of the series of his ' Monograph of Radiated Animals.'

He appears to have had too many occupations to allow him to proceed even with the Part of the work which was to have appeared first ; and my own occupations have since prevented me from resuming the work ; so the plates have been laid aside.

As I see no hopes of the work being now proceeded with by either Prof. Agassiz or myself, and as several of the Plates represent species figures of which have not hitherto been published, I am forced to let them appear as illustrations of the descriptions of certain species, and hope that they may be useful in that humble form.

I may state that I have twice sent copies of the Plates to the late Professor John Müller, the author of a work on the species of *Asteriadæ*, requesting him to inform me whether he had described any of the species, and, if he had, to send me the names which he applied to them ; but though he replied to other portions communication, he made no observation on the Plates, further than to express his admiration of them ; so I suppose that most, if not all, of them were new to him.

The article in the 'Annals and Magazine of Natural History,' vol. vi. p. 165, contains a history of the different arrangements of the family and the genera into which the species have been divided, with the names which have been given to the group, arranged according to the dates of their publication ; and in the introduction to the second paper I gave a review of Professor J. Müller and D. Troschel's paper in the Berlin Academy Reports, and their work on the Starfishes published in 1842.

The specimens described in this Synopsis (Ann. & Mag. Nat. Hist. 1840) were either in the collection of the British Museum or in that of the Zoological Society, which includes the specimens discovered by Mr. Cuming during his residence in South America, and presented by him to the Society.

Those described in the paper in the 'Proceedings of the Zoological Society' for 1847, which is reprinted in the 'Annals and Magazine of Natural History' for the same year, are all contained in the collection of the British Museum. Unfortunately, when the Zoological Society distributed its Museum, several of the species of Starfish which had been brought home by Mr. Cuming and described in my first paper could not be found; and as the time when and how the specimens disappeared could not be discovered, there was no means of tracing the type specimens and procuring them, like other type specimens of that collection, for the British Museum.

The hard parts of these animals, whether they are in the form of *tesseræ*, as in the *Echidua*, or of *ossicula*, as in the *Hypostomata*, or in that of spines, as in either, are evidently the hardening of certain parts of the cellular substance or skin, and these hard parts retain their organization and vitality during the life of the animal; consequently they are not inorganic secretions, like the shells of mollusca, as they have generally been considered, but have far more relation to bones and coral, and like them form a peculiar kind of body, intermediate between shells and the skeletons of vertebrata. "These pieces," as I have observed in the 'Synopsis of the British Museum,' "are formed by the earthy particles being deposited round certain definite spots in the skin; and as they are developed they assume a definite arrangement into certain distinct shapes peculiar to the different kinds. Although these are strongly united together by the skin, and have a kind of organization during the life of the animal, they may easily be separated from each other after death, and then appear like separate bones. This structure allows the animal to increase both the size and the number of the pieces that compose its protecting case as the body grows, and also to repair, by the deposition of fresh calcareous particles on the skin of the healed part, any injury which the animal may have received from external accidents during its life."

This structure is not so easily demonstrated in the internal ossicula of the Starfishes as it is in the external tesseræ of the Sea-eggs and in the spines of both these kinds of animals, as they are often to be found broken and repaired during their growth; and this repair does not take place by any secretion applied to their surface, but by a healing of the part, which leaves a scar on the surface. Nevertheless the entire similarity which exists between the external spines and the internal tubercles at once shows that they are of the same structure; and this is further proved by the examination of the tubercles of those kinds which are in great part exposed on the surface, as is the case with the different kinds of *Pentaceros*, where the development of these hard parts can often be observed during the process of reproducing an arm that has been accidently injured or destroyed.

SYNOPSIS

OF THE

SPECIES OF STARFISH.

Order 1. ASTEROIDA.

Body free, star-shaped, with distinct small ambulacra (or walks) of double pores on the oral surface, from the mouth to the ends of the rays; dorsal wart distinct. These animals have the faculty of reproducing the arms or such parts as may be accidentally broken off; and if an entire arm be separated, provided part of the body be attached to it, other arms are reproduced, and a fresh perfect animal formed. *Gray, Ann. N. H.* 1840, p. 178.

SECT. 1. *The ambulacra with four rows of feet; dorsal wart simple.*

Family I. ASTERIADÆ, *Gray, Syn. Brit. Mus.* 62 ; *Ann. N. H.* 1840, p. 178. Asteracanthion, *Müll. & Trosch. Ast.* 14, 1842.

A. *Back with spines.*

1. ASTERIAS. Skeleton netted, with a single mobile spine at each anastomosis of the ossicula; body covered with more or less prominent elongated mobile spines. *Gray, Ann. N. H.* 1840, p. 178.

a. *Rays 12 or 13, slender, tapering, with small elongated spines.*

1. *Asterias aster.* Rays three times as long as the diameter of the body; back with seven series of spines; the labial spine at the angles of the arms very long. *Gray, Ann. N. H.* 1840, p. 178 ; *Müll. & Trosch. Ast.* 18. Inhab. —— (Brit. Mus.).

b. *Rays 6 or 8, cylindrical.*

2. *Asterias calamaria.* Arms four times as long as the diameter of the body, with seven ridges of spines; the five dorsal ridges equidistant. *Gray, Ann. N. H.* 1840, p.179; *Müll. & Trosch. Ast.* 19. Inhab. Isle of France, New Holland (Brit. Mus.).

c. *Rays 5–8, elongated, subcylindrical, with five or seven series of spines, the two lower series close together and near the ambulacra.*

3. *Asterias tenuispina.* Rays four or five times as long

as the diameter of the body; spines acute. Var. 1. 8-rayed ; var. 2. shorter-rayed : Madeira. *Asterias glacialis,* Grube, 21 ; Gray, Ann. N. H. 1840, p. 179. *Ast. spinosa,* Pennant. *Asterias tenuispina,* Lamk. ii. 561. *A. savaresii,* Chiaje. *Asteracanthion tenuispianua,* Müll. & Trosch. Ast. 11. Inhab. English Coast, Mediterranean.

4. *Asterias rustica.* Rays 6, flat, broad ; spines short, thick, truncated. *Gray, Ann. N. H.* 1840, p. 179. Asterias gelatinosa, *Meyer, Reise,* i. 222? Asteracanthion gelatinosum, *Müll. & Trosch. Ast.* 19. Inhab. Valparaiso, in sandy mud, *H. Cuming, Esq.*

This species has a series of small triangular plates, pierced with a central triangular hole, within the marginal ambulacral spines.

5. *Asterias echinata.* Rays 8, twice as long as the width of the body, five-sided ; central ridge of spines interrupted. *Gray, Ann. N. H.* 1840, p. 179; *Müll. & Trosch. Ast.* 19. Valparaiso, on mud, about 4 to 6 fathoms, *H. Cuming, Esq.*

d. *Rays 5, elongate, angular.*

Asterias glacialis, Müller, Prodr. 234; Gmelin, 3162 ; Lamk. ii. 561. *Asterias angulosa,* Müller, Zool. Dam. t. 41 ; E. M. t. 119. f. 1. *Stellonia glacialis,* Nardo. *S. angulosa,* Agassiz. *Asteracanthion glaciale,* Müll. & Trosch. 14 ; Linck, t. 38, 39. Inhab. North Sea.

e. *Rays 5, tapering ; the ambulacral series of spines crowded, as if two- or three-rowed; back netted with a ridge of two or three rows of spines next the ambulacral series, and then a single series of spines.* Gray. Ann. N. H. 1840, p. 179.

6. *Asterias Holsatica.* Rays tapering, nearly three times as long as the width of the body. *Gray, Ann. N. H.* 1840, p. 179 ; *Retz. Ast.* 22 & 26. Asteriasviolacea,*Muller,Z.D.* ii. t. 46. A. glacialis, *John.* Asteracanthion violaceum. *Müll.& Trosch.Ast.* 16. Inhab. Northern Europe. Colour very variable.

7. *Asterias rubens.* Rays broad, more than twice as long as the width of the body, with scattered blunt spines, spinulose at the tip. *Gray, Ann. N. H.* 1841, p. 179. Asterias rubens, *Retz. Vet. Akad. Hand.* iv. 236; *Gmelin,*

B

3161. Stellonia rubens, *Nardo*. Asteracanthion rubens, *Müll. & Trosch. Ast.* 17, 126. Inhab. European ocean. Is not this only the female with eggs of the former?

8. *Asterias Katherinæ.* Rays 6 or rarely 5, nearly three times as long as the width of the body; back with scattered and crowded blunt, rough-tipped spines. *Gray, Ann. N. H.* 1840, p. 179: *Müll. & Trosch. Ast.* 19. Inhab. North America, mouth of the Columbia river, *Lady Katherine Douglas.*

9. *Asterias Wilkinsonii.* Rays 5, nearly three times as long as the width of the body; back with about seven irregular interrupted series of rather blunt rough spines. *Gray, Ann. N. H.* 1840, p. 179; *Müll. & Trosch. Ast.* 19. Inhab. Northern Africa, *Sir J. G. Wilkinson.*

See a. Asteracanthion africanum, *Müll. & Trosch. Ast.* 15; Cape of Good Hope. b. Ast. polaris, *Müll. & Trosch. Ast.* 16; Greenland (B.M.). c. Asterias rosea, *Müller, Z. D.* t. 67; *E. M.* t. 116. f. 2, 3; Asteracanthion roseum, *Müll & Trosch. Ast.* 17; North Sea. d. Ast. Bootis, *Müll. & Trosch. Ast.* 17 (Mus. Paris). e. Ast. striatum, *Lamk.* ii. 564; *Müll. & Trosch. Ast.* 18. f. Ast. Groenlandicus, *Steenstr.*; Greenland (B.M.). g. Ast. problema, *Steenstr.*; Greenland (B.M.). h. Ast. Mülleri, *Sars*; Greenland (B.M.)

f. *Body discoidal, divided at the edge into numerous short tapering rays; the series of spines near the ambulacral series rather crowded, large, and elongated.* HELIASTER, Gray, Ann. N. H. 1840, p. 179.

10. *Asterias helianthus.* Arms 33 or 34, about a quarter the length of the width of the body, with three equidistant series of short blunt spines. *Gray, Ann. N. H.* 1840, p. 180; *Lam.* 20; *E. Méth.* t. 108, 109. Stellonia helianthus, *Agassiz.* Asteracanthion helianthus, *Müll. & Trosch. Ast.* 18. Inhab. Guasco, Chili, *Say*; Valparaiso, *H. Cuming, Esq.*

11. *Asterias Cumingii.* Arms 30 or 31, very short, not one-tenth as long as the diameter of the body, conical, with blunt spines. *Gray, Ann. N. H.* 1840, p. 180. Inhab. Hood's Island, on rocks at spring tide, *H. Cuming, Esq.*

12. *Asterias multiradiata.* Arms 22 or 24, cylindrical, elongated, tapering at the ends, one-third longer than the diameter of the body; the dorsal series of spines rather longer and more compressed. *Gray, Ann. N. H.* 1840, p. 180. Inhab. Hood's Island, *H. Cuming, Esq.*

II. UNIOPHORA. Body rather depressed; rays broad, blunt; skeleton formed of transverse oblong ossicula, each bearing a large unequal-sized subglobular articulated spine, placed in longitudinal series. The ambulacra with two or three series of equal equidistant filiform blunt spines on each side. Dorsal wart convex, complicated. *Gray, Ann. N. H.* 1840, p. 288; *Müll. & Trosch. Ast.* 21.

1. *Uniophora globifera.* Rays short, broad, rounded, with globular tubercles, *Gray,l.c.* p. 288. Inhab. Van Diemen's Land, *Ronald Gunn, Esq.*

B. *Back with mobile tubercles.*

III. MARGARASTER. Skeleton largely reticulated, with smooth conicles between the reticulations, which are sharp-edged and studded with rounded granules; ambulacra bordered with a series of short small spines.

1. *M. graniferus.* Asterias granifera, *Lamk.* ii. 560. Asteracanthion graniferum, *Müll. & Trosch.* 20, t. 1, f. 2? Asterias serrulata, *E. M.* t. 104? Ast. jauthina, *Brandt.* Inhab. Pacific Ocean.

IV. TONIA. Skeleton netted, with a series of crowded small blunt mobile spines on the sides of each ossiculum; ambulacra bordered with a crowded series of subulate spines, and without any triangular pierced pieces within them. *Gray, Ann. N. H.* 1840, p. 180.

1. *Tonia atlantica.* Rays 5, more than twice as long as the width of the body; back with nine series of cross bands. *Gray, Ann. N. H.* 1840, p. 180. Asterias aurantia, *Meyer, Reise,* i. 222. Stichaster striatus, *Müll. & Trosch. Monatsb. Berlin,* 1840. Asteracanthion aurantineum, *Müll. & Trosch. Ast.* 21, t. 1. f. 3, 1842. Inhab. Valparaiso, on rocks at low water, *H. Cuming, Esq.*

See also a. Asterias ochracea, *Brandt,* 69. b. Asteracanthion margatiferum, *Müll. & Trosch. Ast.* 20.

V. MITHRODIA. The rays cylindrical, elongate, spinulose; the skeleton netted, with scattered small rugose spines, and series of large clavate spinulose spines regularly articulated to a broad expanded base on the sides of the arms; ambulacra with very fine long hair-like spines placed in rounded groups, with a series of large spines near them. *Gray, Ann. N. H.* 1840, p. 287.

1. *Mithrodia spinulosa.* Arms five times as long as the width of the body, with a series of large spines on each side. The series of spines next to those on the edge of the ambulacra are sometimes hatchet-shaped. *Gray, Ann. N. H.* 1840, p. 288. Asterias clavigera, *Lam.* ii. 29? Pentadactylosaster reticulatus, *Linck,* t. 6. and 10. f. 16? Asterias reticulata, *Blainv. Man.?* not Linn. nor Lam. Asteracanthion Linckii, *Müll. & Trosch. Ast.* 18.

SECT. 2. *The ambulacra with only two rows of feet.*

Family II. ASTROPECTINIDÆ.

Back flattish, netted, with numerous tubercles, crowned with radiating spines at the tip, called paxilli; vent distinct. Astropectinidæ, Gray, Ann. N. H. 1840, p. 180.

A. *The margin of the rays ciliated, with a series of simple elongated spines, the paxilli or crowned tubercles regularly radiating.* Astropecten, *Linck.* Asterias, *Agassiz.* Stellaria, *Nardo.*

a. *The rays edged with a series of large regular tubercles, which increase in number as the animal grows.*

I. NAURICIA. The ambulacral spines broad and ciliated; two series of tesseræ between the angles of the arms and

3

the mouth beneath. *Gray, Ann. N. H.* 1840, p. 180. Asiatic.

1. *Nauricia pulchella.* Rays 5, half as long as the width of the body, gradually tapering; lower series of marginal tubercles with a series of broad flat spines on the upper margin of each. *Gray, Ann. N. H.* 1840, p. 180; *Seba*, iii. t. 8. f. 7, *a, b* (not good). Inhab. China? Japan?
The *Stellaster sulcatus*, Möbius, l. c. t. 4. f. 1, 2, appears to be from a specimen which has lost its marginal spines.

II. ASTROPECTEN. Ambulacral spines simple, linear, without any tessera between the marginal tubercles near the mouth and angles of the arms. *Gray, Ann. N. H.* 1840, p. 180.

1. *Body pentagonal; rays short.* Gray, l. c. 1840. *Ctenodiscus,* Müll. & Trosch. Ast. 76.

1. *Astropecten polaris,* Gray, Ann. N. H. 1840, p. 180. *Asterias polaris,* Sabine, Append. Parry's Voy. 223, t. 1. f. 2, 3. *Ctenodiscus polaris,* Müll. & Trosch. Ast. 76, t. 5. f. 5, 129. *Ctenodiscus crispatus,* Lutken (B.M.). *Asterias crispata,* Retz. Diss. 17. Inhab. North Sea.
Astropecten corniculatus, Linck, t. 27 & t. 36. f. 63; perhaps a variety of the former; and *Ctenodiscus pygmœus,* Müll. & Trosch. Ast. 76, p. 129, is the young.

2. *Body 5-rayed; arms depressed: the upper series of marginal tubercles broad, rounded or sheleing towards the edge.*

a. *The dorsal tubercles between the angles of the arms on the centre of the back and on the lines down the centre of the arms the largest.* Gray, l. c. 1840. *Archaster,* Müll. & Trosch. Ast. 1842.

2. *Astropecten stellaris,* Gray, Ann. N. H. 1840, p. 181. *Archaster typicus,* Müll. & Trosch. Monatsb. Berl. 1840; Ast. 65, t. 5. f. 2 (B.M.). Inhab. ——.
See *a.* Archaster hesperus, *Müll. & Trosch. Ast.* 66. *b.* Archaster angulatus, *Müll. & Trosch. Ast.* 66.

b. *The dorsal tubercles subequal, with fasciculated spines.* Gray, l. c. 1840. *Astropecten,* Müll. & Trosch. 1841.

† *The oral series of marginal tubercles produced beyond the dorsal ones.*

* *The upper marginal tubercles with a single series of spines at the angle of the base of the rays, and with another series at the end of the rays, which together make a double series near the base of the rays.*

3. *Astropecten duplicatus.* Rays three times as long as the diameter of the body, slender; marginal spines elongated, depressed, linear. *Gray, Ann. N. H.* 1840, p. 181. A. Brasiliensis, *Müll. & Trosch. Ast.* 68? Inhab. St. Vincent's, *Rev. L. Guilding.*

4. *Astropecten aurantiacus.* Rays three times as long as

the diameter of the body, slender; marginal spines subulate, elongated. *Gray, Ann. N. H.* 1840, p. 181; *Müll. & Trosch. Ast.* 67. Asterias aurantiaca, *Linn.*; *Philippi, Wiegm. Arch.* 1837, p. 193; *Linck*, t. 3. f. 6, t. 6. f. 6; *E. M.* t. 110. f. 23. Inhab. Mediterranean.

5. *Astropecten stellatus.* Rays more than twice as long as the width of the body. The central area of the arms is about as wide as one series of the marginal tubercles. *Gray, Ann. N. H.* 1840, p. 181? Astropecten Valenciemi, *Müll. & Trosch. Ast.* 68. Inhab. Coast of South America?
See *a.* Astropecten Tiedemanni, *Müll. & Trosch. Ast.* 69 (Mus. Vienna).

** *The upper series of marginal tubercles with a continued single series of spines on the angle of the arms.*

6. *Astropecten armatus.* Rays elongate, regularly tapering; upper marginal tubercles narrow, with a continued series of erect, elongated, subulate spines.
Var. *pulcher.* The under series of marginal tubercles not produced, and the spines more slender. *Gray, Ann. N. H.* 1840, p. 181. Inhab. Puerto Portrero, South America, on sandy bottoms, 9 fathoms, *H. Cuming, Esq.* (var.).

7. *Astropecten echinatus.* Rays rather more than twice as long as the width of the body; upper series of spines large; lower series depressed, acute. *Gray, Ann. N. H.* 1840, p. 181; *Linck*, 29, t. 8. f. 12, 12.
See *a.* Astropecten bispinosus, *Müll. & Trosch. Ast.* 64. *b.* Asterias bispinosa, *Otto, Nor. Acta Leop.* xi. p. 285, t. 39.

*** *The upper series of marginal tubercles spineless, the lower series much produced.*

8. *Astropecten marginatus.* Rays nearly three times as long as the width of the body; lower marginal tubercles linear, depressed. *Gray, Ann. N. H.* 1840, p. 181. *Astropecten fimbriatus,* Linck, is probably this species with the marginal spines lost.

9. *Astropecten regalis.* Rays one-fourth longer than the diameter of the body, broad, tapering; spines broad, blunt, depressed. *Gray, Ann. N. H.* 1840, p. 181. Inhab. St. Blas, *H. Cuming, Esq.*
Like *A. marginatus,* but the arms are shorter and broader.

**** *The upper series of marginal tubercles with two series of spines at the base and one along the edge of the arms.*

10. *Astropecten erinaceus.* Arms gradually tapering, twice as long as the width of the body; upper marginal tubercles rather narrow, with a series of small short spines, and a series of six or eight larger ones. *Gray, Ann. N. H.* 1840, p. 182. "St. Elena, sandy mud, 6 fathoms," *H. Cuming, Esq.*

†† *The under or oral series of marginal tubercles rounded, and not produced beyond the dorsal ones.*

* *The upper series of marginal tubercles with a series of short spines.*

11. *Astropecten Mauritianus.* Rays broad; lower spines

B 2

4

broad, strap-shaped. *Gray, Ann. N. H.* 1840, p. 182.
Archaster angulatus, *Müll. & Trosch. Ast.* 66. Inhab. Isle
of France.

** *Upper series of marginal tubercles spineless.*

12. *Astropecten mesodiscus.* Rays elongate, slender, tapering; upper marginal tubercles narrow, with two series
of short small tubercles like granules, one on each of the
margins; lower spines broad, elongate. *Gray, Ann. N.
H.* 1840, p. 182; *Linck*, 20, t. 4. f. 16. Inhab. ——.

13. *Astropecten gracilis.* Rays elongate, slender, gradually tapering; upper marginal plates rather broad, granular, with fine spines on the suture between them; lower
spines small, blunt, depressed. *Gray, Ann. N. H.* 1840,
p. 282. Inhab. ——. Like the former, but arms narrower.

14. *Astropecten irregularis.* Rays rather broad, tapering; the upper tubercles rather broad, with a series of one
or two scattered tubercular spines near the tip; lower
spines depressed, acute. *Gray, Ann. N. H.* 1840, p. 182;
Linck, 27, t. 6. f. 13. Ast. aurantiaca, *Müller, Z. D. t. 83.*
Ast. Johnstonii, *Chiaje?* Inhab. Pembrokeshire, *Linck.*

15. *Astropecten dubius.* Rays broad, tapering; upper
marginal tubercles rather broad, granular, spineless (?);
lower spines broad, depressed. *Gray, Ann. N. H.* 1840,
p. 182. Inhab. West Indies.

*** *Upper and lower margin spineless, serrated* (?).

16. *Astropecten regularis*, Gray, Ann. N. H. 1840, p. 182;
Linck, 26, t. 8. f. 11. Asterias petalodea, *Retz. Aster.* 16.
n. 14? Inhab. ——. I have never seen this species.

3. *Body 5-rayed, the arms high, narrow; upper marginal
tubercles very narrow and erect; the line of dorsal
tubercles down the centre of the arms the largest.*
Astropus, *Gray, Ann. N. H.* 1840, p. 182.

17. *Astropecten longipes.* Rays long and narrow; the
upper marginal tubercles minutely granular, and one or
two often furnished with a short broad conical
spine; lower with a broad depressed blunt erect addressed
spine. Monstrosity 4-rayed. *Gray, Ann. N. H.* 1840, p. 182.
Inhab. "Isle of France," *Leach.*

See *a.* Astropecten polyacanthus, *Müll. & Trosch.* 69,
t. 5. f. 3; Red Sea. *b.* Asterias platyacanthus, *Phil. Wiegm.
Arch.* 1837, p. 193; Astropecten platyacanthus, *Müll. &
Trosch. Ast.* 70; Mediterranean. *c.* Astropecten hystrix,
Müll. & Trosch. Ast. 70; Ceylon. *d.* Astropecten armatus,
Müll. & Trosch. Ast. 71; Japan. *e.* Astropecten scoparius,
Müll. & Trosch. Ast. 71 (Mus. Paris). *f.* Astropecten Hemprichii, *Müll. & Trosch. Ast.* 71; Red Sea. *g.* Asterias
articulatus, *Say, Journ. Acad, Phil.* i. 144. Astropecten
articulatus, *Müll. & Trosch.* 72; Florida. *h.* Asterias Johnstoni, *Chiaje, Mem.* t. 18. f. 4; *Philippi*; Astropecten Johnstoni, *Müll. & Trosch. Ast.* 72; Mediterranean. *i.* Astropecten serratus, *Müll. & Trosch. Ast.* 72; Mediterranean.
j. Asterias spinulosa, *Philippi, Wiegm. Arch.* iii. p. 193;
Astropecten spinulosus, *Müll. & Trosch. Ast.* 72. *k.* Astropecten Japonicus, *Müll. & Trosch. Ast.* 73; Japan. *l.* As-

tropecten hispidus, *Müll. & Trosch. Ast.* 73 (Mus. Leyden).
m. Astropecten longispinus, *Müll. & Trosch. Ast.* 73. *n.*
Asterias platyacantha, *Chiaje, Mem.* t. 18. f. 1, 3; A. aran-
ciaca, *Johnst. Mag. N. H.* ix. 289, f. 43; Astropecten
platyacanthus, *Müll. & Trosch. Ast.* 74. *o.* Asterias subinermis, *Phillipi, Wiegm. Arch.* iii. 193. *p.* Astropecten
subinermis, *Müll. & Trosch. Ast.* 74; Sicily. *q.* Astropecten
marginatus, *Müll. & Trosch. Ast.* 75 (Mus. Paris). *r.* Astropecten Schœnleinii, *Müll. & Trosch. Ast.* 78; Goré. *s.*
Astropecten granulatus, *Müll. & Trosch. Ast.* 75 (Mus.
Leyden).

b. *The rays without any large tubercles on the margin.*

III. LUIDIA. Margin of the 5 flat rays erect: the dorsal
surface crowded with regular paxilli. *Gray, Ann. N. H.*
1840, p. 183.

1. *Luidia fragilissima*, Forbes in Wern. Trans. 1839,
14; Wern. Mem. viii. 128; Müll. & Trosch. Ast. 77;
Gray, Ann. N. H. 1840, p. 183. Asterias rubens, Johnston
in Mag. N. H. 144, f. 20. Inhab. North Sea.

2. *Luidia Savignii*, Gray, Ann. N. H. 1840, p. 183;
Müll. & Trosch. Ast. 77. *Asterias Savignii*, Audouin in
Savigny, Egypt, Echinod. t. 3. Inhab. Red Sea.

3. *Luidia ciliaris*, Gray, Ann. N. H. 1840, p. 183. *Asterias ciliaris*, Philippi in Wiegm. Arch. 1837, p. 19. Inhab. Sicily.

See *a.* Luidia maculata, *Müll. & Trosch. Ast.* 77; Japan.
b. Asterias Senegalensis, *Lamk.* ii. 567; *E. M.* t. 121;
Luidia Senegalensis, *Müll. & Trosch. Ast.* 78, t. 5. f. 4;
Senegal, Brazils? *c.* Luidia clathrata, *Say*; West Indies
(B.M.). *d.* Luidia tessellata, *Lütken*; West coast, Central
America (B.M.).

IV. PETALASTER. Margin of the rays shelving; the
dorsal surface with equal paxilli placed in longitudinal and
cross series. *Gray, Ann. N. H.* 1840, p. 183. Chætaster,
Müll. & Trosch. Ast. 27, 1842.

1. *Petalaster Hardwickii.* Rays elongated, rather slender,
tapering at the end; the dorsal tubercles with small truncated spines, and a distinct series of rudimentary spines.
Gray, Ann. N. H. 1840, p. 183. Inhab. Indian Ocean.

2. *Petalaster Columbia.* Rays elongated, slender, gradually tapering; tubercles short, with crowded groups of
rather large acute spines, and a fringe of very fine radiating
ones. *Gray, Ann. N. H.* 1840, p. 183. Inhab. St. Blas,
H. Cuming, Esq.

See *a.* Asterias longipes, *Retz. Dis.* 20; Asterias subulata, *Lamk.* ii. 568; *Chiaje, Mem.* t. 21. f. 5, 6; Chætaster
subulatus, *Müll. & Trosch. Ast.* 27, t. 2. f. 1, 127; Mediterranean. *b.* Chætaster Hermanni, *Müll. & Trosch. Ast.*
27. *c.* Chætaster Troschelii, *Müll. & Trosch. Ast.* 28.

B. *The margin of the rays not edged with large tubercles,
simple, or ciliated, with short broad spines bearing
tubercles; vent none.* Allied to *Cribellina*, p. 12.

V. SOLASTER. The rays many, with two series of broad
spines bearing tubercles near the ambulacra. *Gray, Ann.*

N. H. 1840, p. 183; *Forbes, Wern. Trans.* viii. 121; *Müll. & Trosch. Ast.* 26; Crossaster, *Müll. & Trosch.*

a. *Body 8- or 9-rayed, closely reticulated, rays rounded, ventricose below, tapering at the tip, with a second row of compressed tubercles on the underside of the arms near the ambulacral series.* Eudora, Gray.

1. *Solaster endeca*, Forbes, Wern. Trans. viii. 121; Gray, Ann. N. H. 1840, p. 183; Müll. & Trosch. Ast. 26; *Asterias endeca*, Linn. *Ast. aspersa*, Müller. Inhab. European Seas.

b. *Body 10- or 12-rayed, loosely reticulated; the rays depressed, with a series of large compressed tubercles crowned with a bunch of spines edging the oral ridge.* Polyaster, Gray.

2. *Solaster papposus*, Forbes, Wern. Trans. viii. t. 121; Gray, Ann. N. H. 1840, p. 183; Müll. & Trosch. Ast. 26. *Asterias papposa*, Linn. *Ast. stellata*, Retz. Inhab. European Seas.

VI. HENRICIA. The rays 5, rounded, tapering, with rounded tubercles near the ambulacra; the dorsal wart obscure, few-rayed, often hidden with small spines. *Gray, Ann. N. H.* 1840, p. 184. Linckia, *Forbes* not *Nardo.* Echinaster, sp., *Müll. & Trosch.*

1. *Henricia oculata.* Rays 5, closely reticulated with small spines. *Gray, Ann. N. H.* 1840, p. 184. Asterias oculata, *Penn.* Asterias seposita, *Retz. Dis.* 21. Echinaster oculatus, *Müll. & Trosch. Ast.* 24; *Linck*, t. 36. f. 62. Linckia oculata, *Forbes, Wern. Trans.* viii. 120, t. 3. f. 8.

See *a.* Echinaster Eschrichtius, *Müll. & Trosch. Ast.* 25; Greenland.

Family III. PENTACEROTIDÆ.

The body supported by roundish or elongated pieces, covered with a smooth or granular skin, pierced with minute pores between the tubercles; vent none. *Gray, Syn. Brit. Mus.* 1840; *Ann. N. H.* 1840, p. 275.

A. **Pentacerotina.** *Body pentagonal or suborbicular; rays short; dorsal wart single; the ambulacra edged with a series of small spines divided into rounded groups.* Gray, Ann. N. H. 1840, p. 275.

a. *The ambulacra with a single series of large spines near the edge.*

* *Body suborbicular, convex above and below; covered above and below with granules, and scattered conical tubercles.*

I. CULCITA. This genus chiefly differs from *Randasia* and *Pentaceros* in having no upper series of marginal ossicula. It agrees with *Randasia* in the back being nearly flat. *Gray, P. Z. S.* 1847, p. 74; *Agassiz, Prodr.* 25; *Müll. & Trosch. Ast.* 37.

1. *Culcita Schmideliana.* Body subcircular, flat above when dry (very convex subglobose when alive). The back coriaceous, without any apparent reticulations, covered with scattered, small, conical spines. The oral surface rather convex (when dry), closely and minutely granular, and with larger conical tubercles; those near the ambulacra and oral angles much the largest and ovate. *Gray, Ann. N. H.* 1840, p. 276; *P. Z. S.* 1847, p. 74. A. Schmideliana, *Retz. Dis.*; *Schmidel's Naturf.* xvi. t 1 (good). A discoidea, *Lam.* A. placenta, *Lamk.* Culcita discoidea, *Agassiz, Müll. & Trosch. Ast.* 37. Inhab. Lord Hood's Island, on reefs, *H. Cuming, Esq.* Bright orange when alive, when in the water very convex.

There are distinct indications of the lower marginal ossicula in this species, but they and the ossicula of the oral surface are not sufficiently large and close to force the dry specimen to assume the pentangular form of the following species.

2. *Culcita pentangularis* (T. 2. f. 2). Body pentangular; back flat when dry, convex beneath, minutely and closely granulated, with obscure reticulations, the reticulations armed with small conical tubercles; the interspaces closely and minutely porous. The oral surface protected by distinct well-defined ossicula, defining the lower edge of the margin, covered with close and minute granules and larger round-topped tubercles, those near the ambulacra and the oral angles being largest and highest. *Gray, P. Z. S.* 1847, p. 74. Inhab. Reef of Oonaga.

This species is very distinct from the former, and forms the passage to the genus *Randasia*; but there is a series of concave, minutely porous spaces in place of the upper marginal plates.

See *a.* Culcita coriacea, *Müll. & Trosch. Ast.* 38; *E. M.* t. 97. f. 3. *b.* Culcita Novæ-Guineæ, *Müll. & Trosch. Ast.* 38. *c.* Culcita crex, *Müll. & Trosch. Ast.* 34.

II. ASTERODISCUS. Body pentagonal, coriaceous, depressed, covered with numerous close, flat-topped, unequal, small tubercles; back convex; dorsal wart roundish, subcentral; arms short, rounded, with a pair of large convex kidney-shaped ossicula on each side of the tip above. Margin simple, rounded, beneath concave; ambulacra with a series of short linear spines, placed in groups of four or five, each group on a separate ossiculum, and with two series of larger, blunt, club-shaped spines on the outside of the ambulacral spines. The young specimens have indistinct inferior marginal ossicula. *Gray, P. Z. S.* 1847, p. 78.

1. *Asterodiscus elegans* (T. 12. f. 1, 2). Pale brown when dry; tubercles of the back unequal, the larger ones truncated, those nearest the mouth on the underside larger, club-shaped, rather crowded. *Gray, P. Z. S.* 1847, p. 78; *Ann. N. H.* 1847, p. 196. Inhab. —— (Brit. Mus.).

** *Body pentagonal, formed of curiously shaped, regularly arranged, externally granular ossicula.* Gray, l. c. 1840. Oreaster, Müll. & Trosch. Ast. 44.

III. PENTACEROS. Body convex above, margin with two rows of large spine-bearing tesseræ. *Gray, Ann. N. H.* 1840, p. 276.

6

a. *Back formed of irregular elongated ossicula, apparently reticulated; the spines with enlarged bases, interspaces closely punctured.*

1. *Pentaceros grandis.* Arms very broad, as wide as long at the base, only half as long as the width of the body. *Gray. Ann. N. H.* 1840, p. 276; *Seba,* t. 8. f. 1. Diam. 17". Inhab. ——.

2. *Pentaceros reticulatus.* Arms rather broad, nearly as long as the width of the body; back convex. Monstrosity 4-lobed. *Gray, Ann. N. H.* 1840, p. 276; *Rumph. Mus.* t. 15. f. D. Asterias reticulata, *Linn.* Oreaster reticulatus, *Müll. & Trosch. Ast.* 45, t.3. f. 2. Inhab. West Indies, Barbadoes, *Ralph Green, Esq.*

3. *Pentaceros gibbus.* Arms rather shorter than the width of the body; back depressed. *Gray, Ann. N. H.* 1840, p. 276; *Linck,* t. 23. f. 36; *Seba,* iii. t. 7. f. 1. Inhab. West Indies and St. Vincent's, *Rev. L. Guilding.*

See *a.* Pentaceros lentiginosus, *Linck,* 25. t. 41, 42. f. 72. *b.* Ast. pentacyphus, *Retz.,* with smaller spines and a nearly spineless margin. *c.* Pentaceros horridus, *Linck,* t. 25. f. 40.

4. *Pentaceros Cumingii.* The arms rather narrow, nearly as long as the diameter of the body; marginal spines few, small; back rather depressed, with conical protuberances bearing small spines. Diam. 2"6; *Gray, Ann. N. H.* 1840, p.276. Inhab. Punto Santa Elena. Rocky ground 12 or 18 fathoms, *H. Cuming, Esq.*

Perhaps the young of a much larger species.

5. *Pentaceros nodosus.* Arms rather narrow, nearly as long as the width of the body, with a single series of blunt tubercles; back rather depressed, with a large tubercle on each angle of the centre. *Gray, Ann. N. H.* 1840, p. 276; *Linck,* t. 26. f. 41. Ast. nodosa *a, Lamk.; Müll. & Trosch. Ast.* 48. Inhab. Isle of France, *Dr. W. E. Leach.*

In Linck's figure the spines are rather larger than in our specimens of nearly the same size.

6. *Pentaceros Chinensis.* Rays elongated, nearly as long as the width of the body, with small blunt marginal tubercles; back high, with four or five small central tubercles, and a very large blunt tubercle at each angle. *Gray, Ann. N. H.* 1840, p. 276. Inhab. China. *J. Reeves, Esq.*

The central dorsal series of tesseræ are not armed with spines; are they so in larger specimens?

7. *Pentaceros Franklinii* (T. 10). Rays elongate, as long as the width of the body, with a dorsal series of broad blunt tubercles; back high, with very large spines at each angle; margin not armed. *Gray, Ann. N. H.* 1840, p. 277. Var. 1. With one or two conical tubercles on each side of the tubercles, near the one at the angle of the central dorsal disk. Inhab. Coast of New Holland, *G. Bennett, Esq.*

See *a.* Pentaceros turritus, *Linck,* t. 22, 23. f. 3. Like the former, but the back is more spinous, and the spines are not so large.

8. *Pentaceros muricatus.* Arms elongated, nearly as long as the width of the body, with a dorsal series of large, and

with two or three large conical spines near the tips; back rather high, spinous. *Gray, Ann. N. H.* 1840, p. 277; *Linck,* t. 7. f. 8. Ast. Linckii, *Blainv.* A. nodosa, *Lam. Seba,* iii. t. 7. f. 3. Oreaster turritus, *Müll. & Trosch. Ast.* 47. Inhab. ——. (Brit. Mus.)

9. *Pentaceros granulosus* (T. 6. f. 3). Five-rayed; rays as long as the diameter of the disk, rounded at the tip; back rather convex; ossicula convex, rounded, all covered with close rounded granules, the two or three central ones on the top of each ossiculum being larger, those on the middle of the back largest and subtubercular; the marginal ossicula convex, rounded. *Gray, P. Z. S.* 1847, p. 75. Inhab. Western Australia.

Young? Arms more slender, and the lower marginal ossicula near the tips of the arms each with a group of two or three spines, the one nearest the tip largest. The dorsal surface of this species is furnished with abundance of *pedicellaria,* one arising from each hole between the ossicula.

10. *Pentaceros modestus* (T. 9). Arms rather depressed, broad, not quite as long as the diameter of the body, with six or seven large convex tubercles along the middle of the upper surface; the hinder tubercle with a smaller tubercle on each side of it, forming an irregular ring of tubercles exterior to the dorsal ones; back with a circular series of five large ovate blunt spines. Inhab. —— ?

See *a.* Oreaster affinis, *Müll. & Trosch. Ast.* 46. *b.* Oreaster Chinensis, *Müll. & Trosch. Ast.* 46. *c.* Oreaster tuberculatus, *Müll. & Trosch. Ast.* 46. *d.* Asterias mammillata, *And. Descr. Egypt, Echin.* 209, t. 5; Oreaster mammillatus, *Müll. & Trosch. Ast.* 48. *e.* Oreaster verrucosus, *Müll. & Trosch. Ast.* 49. *f.* Asterias stellata, *Mus. Tessin.* 114, t. 9. f. 2; Oreaster clavatus, *Müll. & Trosch. Ast.* 49; *Seba,* t. 6. f. 1, 2, t. 5. f. 7, 8. *g.* Oreaster carinatus, *Müll. & Trosch. Ast.* 49. *h.* Asterias obtusata, *Lamk. E. M.* t. 103; Oreaster obturatus, *Müll. & Trosch. Ast.* 56. *i.* Asterias obtusangula, *Lamk.* ii. 556; Oreaster obtusangula, *Müll. & Trosch. Ast.* 51. *j.* Oreaster regulus. *Müll. & Trosch. Ast.* 51. *k.* Oreaster orientalis, *Müll. & Trosch. Ast.* 827; China. *l.* Oreaster gigas, *Lutken;* West Indies (B.M.).

b. *Back formed of irregular flat-topped ossicula, placed in rows so as to appear nearly tessellated; arms elongated, rather narrow.*

11. *Pentaceros nodosus.* Arms with a double series of hemispherical tubercles; back rather depressed; marginal ossicula unequal, lower ones with small blunt conical spines. *Gray, Ann. N. H.* 1840, p. 172. Asterias nodosa, *Gmelin* (part.), *Seba,* iii. t. 8. f. 11, 12 (t. 5. f. 11, 12, without spines on the margin?). Oreaster nodosus, *Müll. & Trosch. Ast.* 52. Inhab. Isle of France, *W. E. Leach, M.D.*

c. *Back formed of regular rounded ossicula, placed in rows, rather low.*

12. *Pentaceros aculeatus.* With three ridges of small spine-bearing tubercles; back rather depressed, with three small spines at the angles; marginal ossicula rounded, with

conical tubercles. *Gray, Ann. N. H.* 1840, p. 277; *Seba,* iii. t. 5. f. 5, 6. Oreaster aculeatus, *Müll. & Trosch. Ast.* 50. Var.? or younger? Spine-bearing ossicula further apart, with the skin and granulations worn off and bleached, *Seba,* iii. t. 7. f. 1. Inhab. West Indies, St. Vincent's, *Rev. J. Guilding.*

See also *Pentaceros spinosus.* Ast. nodosa (part.), *Gmel. Seba,* iii. t. 5. 7, 8, and var., *Seba,* iii. t. 7. f. 1, 2. Ossicula oblong, with two or three small conical tubercles.

d. *Back regularly convex, formed of flat granular ossicula, with a blunt mobile spine on the centre of each ossiculum below; arms short, broad.* Nidorellia, *Gray, Ann. N. H.* 1840, p. 277.

13. *Pentaceros armatus* (T. 14). Arms short, broad, the lower marginal and the last three upper marginal plates at the top with short blunt spines; back convex, with central and lateral groups and a series of spines down each arm. Young more convex; spines shorter, blunter, and fewer. Younger not so convex, without any marginal spines, and only indications of them on the back. *Gray, Ann. N. H.* 1840, p. 277. Oreaster armatus, *Müll. & Trosch. Ast.* 52. Gonodiscus conifer, *Möbius, Abhandl.* iv. 10, t.3.f. 5, 6, 1860. Inhab. Punto Santa Elena. Rock ground, 12 to 15 fathoms, *H. Cuming, Esq.*

IV. STELLASTER. Body depressed, covered with large, flat, regular, six-sided plates; margin with two rows of large tesserae; the lower rows with a series of compressed mobile spines. *Gray, Ann. N. H.* 1840, p. 277; *Müll. & Trosch. Ast.* 62, 1842.

1. *Stellaster Childreni* (T. 7. f. 2). Back convex, with one or two blunt tubercles on the angles of the centre; arms three-quarters the length of the width of the body, narrow, attenuated to a blunt recurved tip. Young, back without any tubercles. *Gray, Ann. N. H.* 1840, p. 278; *Müll. & Trosch. Ast.* 62, 128, t. 4. f. 3, t. 6. f. 5. Inhab. China or Japan?

See a. Asterias equestris,*Retz.Descr.*12; Stellastercquestris, *Müll. & Trosch. Ast.* 62; Ocean (Mus. Lund). b. Stellaster gracilis, *Möbius, Abhandl.* iv. 1860, ii. t. 11. f. 3, 4, which is very nearly allied, if not the same in a rather different state. This state has also been described by Dujardin under the name of *Astrozonium Souleyetii.*

2. *Stellaster Belcheri* (T. 7. f. 1). Back convex, with two or three large conical tubercles on the line extending to the centre of the arms. Arms slender, tapering, rather longer than the diameter of the disk. *Gray, P. Z. S.* 1847, p. 76. Inhab. Amboina or New Guinea.

This species is intermediate between *S. Childreni* and *S. Incei,* having the white colour and the slender arms of the former, and the convex back and tubercles of the latter; but the tubercles are larger and fewer, and the arms are more slender, having only a single series of plates between the marginal ones. There are two specimens in spirits, and one dry, in the British Museum collection.

3. *Stellaster Incei* (T. 5. f. 1). Purplish, minutely granular; back with scattered, conical, convex tubercles, those

down the centres of the arms largest. The lower marginal plates are flattish. *Gray, P. Z. S.* 1847, p. 76. Inhab. North Australia.

This species is very like *Stellaster Childreni,* Gray, Ann. & Mag. Nat. Hist. 1840, p. 278; Müller, Aster. 62, 128, t. 4. f. 3; *Asterias equestris,* Retzius, Diss. 12; but it is purplish when dry; the back is tubercular; the whole surface is minutely granular; while the Japanese species is always white, the back smooth, and the granules of the surface are so minute and thin that they are very easily eroded, and the lower marginal plates are more convex and the central ones much larger than the others.

V. DORIGONA. Body depressed, 5-rayed, smooth; the dorsal and oral disk covered with many smooth, flat, polygonal squares; the marginal ossicules without any mobile spine.

1. *Dorigona Reevesii* (T. 7. f. 3). Inhab. China or Japan; common in boxes of insects brought from China and Japan. See a. Dorigona longimana= Astrogonium longimanum, *Möbius, Abhandl.* iv. (1860) 7, t. 1. f. 5, 6.

VI. COMPTONIA. Body depressed, spinous (?); dorsal and oral disk covered with very small flat plates; marginal ossicula large, without any mobile spines. *Gray, Ann. N. H.* 1840, p. 278.

1. *Comptonia elegans.* The fossil genus *Corlaster,* Agassiz, from Maestricht, appears to be most nearly allied to this genus; but the plates of the oral disk (which alone are known) appear to be linear longitudinal. *Gray, Ann. N. H.* 1840, p. 278. Fossil. Black Down.

VII. CALLIDERMA. Body flat, five-sided; rays rather elongated; attenuated end only formed of the marginal plates. Ossicula all minutely granulated; the dorsal ossicula flat-tipped, six-sided, some with a larger, globular central tubercle-like granule. The marginal ossicula broad, gradually becoming smaller near the tip, short-edged, minutely granular, those of the upper and lower series alternating; the upper ones with some indistinct spines on the margin, the lower ones with scattered mobile spines on the oral surface. The ossicula of the oral surface three-, four-, or six-sided, granular, with one (rarely two) central, compressed, acute, mobile spine. The ambulacral spines very small, close, fourteen or sixteen on each ossiculum, forming a rounded group, with two or three series of large, scattered, mobile, acute spines on the outer side. *Gray, P. Z. S.* 1847, p. 76.

This genus resembles *Stellaster,* but differs from it in the oral surface being furnished with scattered spines. There is a fossil species, very like the one here described, found in the chalk, and figured in Mr. Dixon's work on the fossils of Worthing, which I propose to call *Calliderma Dixonii.* There are probably several other fossil species from the same locality; they have been referred to the to the genus *Tosia,* but the ossicula are granular and the oral surface spinous.

1. *Calliderma Emma* (T. 15). Flat, pentangular, the sides concave, the arms elongated, produced, tapering to a line

point, about two-thirds the length of the diameter of the disk. The dorsal ossicula six-sided, regular, flat-topped, covered with minute roundish granules: the central granules of the central ossicula and those down the centre of the arms larger, globular, tubercle-like. The margins sharp-edged, concave in the centre; the ossicula of the upper and lower series alternating, minutely granular, with one or two larger subspinose granules on the middle of the upper margin. Marginal ossicula about fifty on each surface on each side, the lower series with scattered, acute, compressed spines on their oral side. The ossicula of the oral side four- or six-sided, rather irregular, minutely granular, each armed with a central, compressed, acute, mobile spine. *Gray. P. Z. S.* 1847, p. 77; *Ann. N. H.* 1847, p. 198. Inhab. ——?

This species most nearly resembles a fossil found in the chalk, which has hitherto been referred to the genus *Tosia*, and figured in Mr. Dixon's work on the fossils of Worthing. I have named this fine species in compliment to my daughter Mrs. J. P. G. Smith, who before her marriage commenced a series of plates to illustrate a monograph of this genus.

*** *Body pentagonal, formed of variously shaped, rather rough ossicula sunk into a naked skin, with a single series of spine-bearing tubercles.* Gray, l. c. 1840. *Asteropsis*, Müll. & Trosch. 62, 1842.

VIII. GYMNASTERIA, *Gray, Ann. N. H.* 1840, p. 278.

1. *Gymnasteria spinosa.* Rays triangular, tapering, about one quarter longer than the width of the body, with a dorsal series of conical cylindrical tubercles. Young with a few spines on the margins and back of the arms. Allied to *Porania. Gray, Ann. N. H.* 1840, p. 278. *Asterias carinifera*, Lamk. ii. 556. *Asteropsis carinifera* Müll. & Trosch. 63, t. 3. f. 4. Inhab. Panama, fine sand, 16 fathoms, *H. Cuming, Esq.*

2. *Gymnasteria inermis.* Rays rapidly tapering, convex above, without any spines. *Gray, Ann. N. H.* 1840, p. 278. Inhab. Panama, in fine sand, 10 fathoms. Half the size of the young spined specimens of the former species.
See a. Asteropsis etenacantha, *Müll. & Trosch. Ast.* 64.

b. *The ambulacra with two series of larger spines near the edge: body depressed: back flat.*

* *The ossicula granulated, sunk in the skin, often spine-bearing.* Gray, l.c. 1840. *Goniodiscus*, Müll. & Trosch. Ast. 57, 1842.

IX. PAULIA, *Gray.* Body 5-rayed, formed of flat granulated spine-bearing irregular ossicula on the disk and margin, without any two-lipped pores. *Gray, Ann. N. H.* 1840, p. 278.

1. *Paulia horrida.* Chestnut-brown; spines acute.
Var. Smaller; arms as long as the width of the body, rather tapering; spines smaller, blunt, rounded at the tip; back more closely granulated. *Gray, Ann. N. H.* 1840, p.278; *Müll.&Trosch.Ast.*64. Inhab. Punta Santa Elena. Rocky ground, 12 to 18 fathoms, *H. Cuming, Esq.*

X. RANDASIA. Body pentagonal, with a tubercular skin above, and large granular plates beneath and on the margin, without any two-lipped slits, but with one or two small pores near the oral angles beneath, where the tubercles are rubbed off. Allied to *Culcita. Gray, Ann. N. H.* 1840, p. 278.

Body pentagonal, depressed, minutely granular; back nearly flat, minutely granular, reticulated; reticulations rather tubercular; interspaces sunken (when dry) and covered with very minute close perforations. Dorsal tubercles roundish, single, subcentral. Margins furnished with an upper and lower series of oblong ossicula; the upper ones narrower internally, with a central series of tubercles; the lower ones oblong, close together, and convex. The oral surface protected by close, regular, squarish, convex ossicula, covered with short crowded granules. The ambulacral spines in rounded groups; the series of tubercles nearest the ambulacra larger, crowded, and placed in groups of three or five, and those in the oral angles largest and flat-topped. *Gray, P. Z. S.* 1847, p. 74.
This genus differs from *Pentaceros* in the back being flat, elevated, and not angular; it is in several respects intermediate between *Culcita* and *Pentaceros*.

1. *Randasia granulata* (T. 2. f. 1). Body five-sided; back minutely granular, with roundish convex subconical tubercules in the reticulations; the marginal plates fourteen on each side, the upper ones with a central series of tubercles. *Gray, P. Z. S.* 1847, p. 75; *Ann. N .H.* 1847, p. 196. Inhab. Reefs of Attagor, Torres Straits.
There are two specimens of this species in the British Museum, one in a very bad state.

2. *Randasia spinulosa* (T. 12. f. 3). Body five-sided; back and upper marginal plates covered with numerous small, conical, acute spines, without any larger tubercles; the upper marginal plates indistinct. *Gray, P. Z. S.* 1847, p.75. Inhab. ——?
This species is very like the former in shape, size, and appearance, but is very easily known from it by the numerous mobile acute spines with which the back and upper part of the margin are covered, appearing to take the places of the small granulations, and by the absence of the tubercles on the elevated ribs of the back.

3. *Randasia Luzonica.* Thick, brown; the tubercles of the underside unequal, the larger ones flat-topped; sides straight. *Gray, Ann. N. H.* 1840, p. 278. Inhab. Island of Luçon, in the Port of Sual, *H. Cuming, Esq.*

XI. ANTHENEA. Body five-rayed, chaffy, with immersed elongated tubercle-bearing ossicula; margin with regular rows of large tesserae; both surfaces (especially the under) scattered with large two-lipped pores. *Gray, Ann. N. H.* 1840, p.279; *P. Z. S.* 1847, p. 77.

† *A very large two-lipped pore on each ossiculus of the oral surface; the back netted and chaffy.*

1. *Anthenea Chinensis.* Back obscurely netted, rather chaffy, with scattered truncated tubercles in rather diverging lines; marginal plates not tubercled; rays broad, half the length of the width of the body. *Gray, Ann. N.*

H. 1840, p. 279. Asterias chinensis, *Gray, Brit. Mus.*
Asterias pentagonula, *Lamk.* ii. 554. Goniodiscus pentagonula, *Müll. & Trosch. Ast.* 57, t. 4. f. 2. Goniaster articulatus, *Agass. Mus. Paris!* Inhab. China, Japan, *J.
Reeves, Esq.*
    See *a.* Seba, iii. t. 6. f. 5, 6 (*Ast. tessellata*, var. A, Lam.).
Similar, but the dorsal tubercles are larger and angular.
b. Goniodiscus scaber, *Möbius, Abhandl.* iv. 1860, t. 3. f. 3,
4, appears to be a nearly allied species.

2. *Anthenea tuberculosa* (T. 4. f. 1). Back obscurely netted, rather chaffy, with scattered, long, flat-backed tubercles. Marginal ossicula with some moderate granules,
the upper ossicula with one or more large flat-topped tubercles on their upper part. *Gray, P. Z. S.* 1847, p. 77;
Ann. N. H. 1847, p. 198. Inhab. Port Essington.
    This species is very like *Anthenea Chinensis,* Gray (*Aste-
rias pentagonula,* Lam.?), but differs from it in being more
convex and netted and more distinctly tubercular, and in
the upper marginal tesserae being armed with tubercles.
Like the Chinese species, all the ossicula, both marginal
and discal, of the oral surface, are furnished with large,
elongated, two-lipped pores.

†† *One or more small two-lipped pores on some of the
    ossicula of the oral surface; the back subtubercular,
    and the ossicula all covered with large roundish gra-
    nules.*

3. *Anthenea granulifera* (T. 5. f. 2). Both surfaces covered
with small roundish granules, the back with rather convex
ossicula; the arms as long as the diameter of the body;
back with one or two scattered tubercles.
    Var. Back with a blunt tubercle on the centre of each
of the ossicula of the middle of the back. *Gray, P. Z. S.*
1847, p. 77; *Ann. N. H.* 1847, p. 198. Inhab. ——?
    This species is easily known from the former by the
smaller granules on the surface, the length of the arms,
and the small size of the two-lipped pores; those of the
dorsal surface are very minute.

XII. HOSEA. Body depressed, flat, 5-rayed, formed of
distinct, hexangular, nearly equal, slightly tubercular ossicula; back with small and beneath with larger two-lipped
slits. *Gray, Ann. H.* 1840, p. 278.

1. *Hosea flavescens.* Arms two-thirds the length of the
width of the body. *Gray, Ann. N. H.* 1840, p. 278. Inhab. ——. Perhaps young.
    See *a.* Asterias granularis, *Retz. in Müller, Zool. Dan.*
t. 92. f. 1, 4; from the North Sea. Gmelin referred
Linck, t. 13. f. 22, t. 23. f. 37 ?, t. 24. f. 39, and t. 27. f. 45
(all *Goniaster tessellatus*) to this species, as he also has
done to *Ast. equestris.*

2. *Hosea spinulosa* (T. 4. f. 2). Body flat, pentagonal;
sides concave; arms not half the length of the diameter of
the body; ossicula large, subequal, six-sided, very mi-
nutely granular. Marginal ossicula | 8 on each side, convex,
deeply separated from each other, with a series of two or
three small, acute, spine-like tubercles in the centre of
each. The ossicula of the oral surface flat, minutely gra-

nular, with small two-lipped pores. *Gray, P. Z. S.* 1847,
p. 78; *Ann. N. H.* 1847, p. 199. Inhab. Indian Ocean;
Philippines.
    This species nearly resembles in shape *Tosia austra-
lis,* but is at once known from that species by the granular
ossicula, the spines on the margin, and the two-lipped
pores beneath; it differs from *Hosea flavescens* in its being
five-sided instead of five-armed, and in having no spines on
the middle of the back.
    See *a.* Goniodiscus Seba, *Müll. & Trosch. Ast.* 58; Astero-
ceras altera, *Seba,* iii. t. 6. f. 7, 8. *b.* Goniodiscus placenta,
Müll. & Trosch. Ast. 59. *c.* Goniodiscus regularis, *Müll.
& Trosch. Ast.* 59; Pentagonaster regularis, *Linck,* t. 13;
E. M. t. 96; *Seba,* iii. t. 8. f. 4. *d.* Asterias pleyadella,
Lamk. ii. 553; *Seba,* iii. t. 6. f. 5, 6. Goniodiscus pley-
adella, *Müll. & Trosch. Ast.* 54. *e.* Asterias cuspidatus,
Lamk. ii. 553. Goniodiscus cuspidatus, *Müll. & Trosch.
Ast.* 60. *f.* Goniodiscus mammillatus, *Müll. & Trosch.
Ast.* 61. *g.* Goniodiscus capella, *Müll. & Trosch. Ast.* 61.

** *The ossicula of the upper and lower surfaces and the
    margin smooth, with a single continued series of uni-
    form granules round each of their edges.* Gray, Ann.
    N. H. 1840, p. 278. Astrogonium, Müll. & Trosch. Ast.
    52, 1842.

XIII. HIPPASTERIA. Body four- or five-sided, formed of
roundish ossicula, with a large truncated central tubercle;
upper and lower surface with two-lipped pores. *Gray,
Syn. Brit. Mus.* 1840; *Ann. N. H.* 1840, p. 279.

1. *Hippasteria Europaea.* Rays 5, broad, nearly half as
long as the width of the body; marginal ossicula with
three blunt tubercles placed in a central cross series.
Gray, Ann. N. H. 1840, p. 279. Asterias equestris, *Penn.
B. Z.* iv. p. 130; *Sow. Brit. Mis.* f. 3. Asterias phry-
giana, *Parelius, O. F. Müller, Gmelin.* Goniaster phry-
gianus, *Müll. & Trosch. Ast.* 52. Inhab. European Ocean.

2. *Hippasteria Johnstoni.* Rays 4, elongated, slightly
tapering; back spinulose, with short truncated spines;
margin with three or four series of elongated tapering
spines. *Gray, Ann. N. H.* 1840, p. 279. Asterias Johnstoni, *Gray, Johnst. Mag. Nat. Hist.* 1836, vi. f. 21, not
Chiaje. Inhab. North of England.

3. *Hippasteria plana.* Pentaceros planus, *Linck,* 21, t. 12.
f. 21, cop. *E. M.* t. 101 (Ast. equestris, *Gmelin & Lamk.*,
which chiefly differs in the arms appearing longer).

4. *Hippasteria cornuta.* Pent. longiorum cornuum, *Linck,*
43, t. 33. f. 53; *E. M.* t. 102; with the arms still longer and
more slender at the end.
    All four are perhaps varieties of one. Gmelin refers to
this species Linck, t. 5. f. 13 (an *Astropecten*), t. 13. f. 22,
t. 23. f. 37, t. 24. f. 39, and t. 27. f. 45 (all *Goniaster tessellatus*).

XIV. CALLIASTER. Body 5-rayed, with flat immersed
ossicula armed with flat-based deciduous conical spines,
and without any two-lipped slits on either surface. *Gray,
Ann. N. H.* 1840, p. 280.

c

1. *Calliaster Childreni* (T. 13), Grey. Back slightly convex, with a centre, a ring, and five radiating lines of small spines; rays slender, tapering, as long as the width of the body; each of the marginal pieces with a central series of three distant spines. *Gray, Ann. N. H.* 1840, p. 280. Inhab. Japan.

XV. ASTROGONIUM. Body pentangular, flat above and below. Back and oral surface protected by triangular ossicula, each covered with numerous erect, cylindrical, truncated tubercles or granules, those of the oral surface longest. Margin strengthened with regular, oblong, four-sided ossicula, covered with small regular granules, except on the most convex part of their centres; those of the upper and lower series opposite each other. Dorsal wart single. Ambulacra with cylindrical truncated spines, in groups of four on each ossiculum, of equal size, not forming a rounded group, and with a series of similar, rather larger spines on their sides, and a series of small ossicula with terminal granules on their outer sides. Bilabiate slits on either surface. *Gray, P. Z. S.* 1847, p. 78. Astrogonium, sp., *Müll. & Trosch. Ast.*

A. *Body flat, five-sided; granules short; ossicula flat-topped, not tubercled.*

1. *Astrogonium granulare* (T. 1. f. 4). Pentagonal; sides rather concave. Back bright crimson; oral surface yellowish; marginal ossicula oblong, 1¼ on each side, rather convex, except at the most convex part of the upper and lower surface, with very minute granules. Dorsal ossicula hexagonal, flat-topped, with short flat-topped granules; ossicula of oral surface similar, but granules longer. *Gray, P. Z. S.* 1847, p. 79. Asterias granularis, *Retz. Dis.* 10; *Müller, Zool. Dan.* t. 92. f. 1; *Müll. & Trosch. Ast.* 57. Inhab. North Sea (British Museum).

This species is very like *Tosia australis*, but is at once known from it by the granules covering the greater part of the surface of the marginal ossicula.

2. *Astrogonium miliare* (T. 1. f. 3). Flat, dark red, pentangular; rays rounded at the end, about one-third the length of the diameter of the disk. Margin rounded, ossicula 2/3 or 3/4 on each side, covered with uniform, close granules. Dorsal ossicula rather convex, covered with uniform granules. *Gray, P. Z. S.* 1847, p. 80. Inhab. New Zealand.

Like *A. granulare* in form, but the margin is round, and the marginal plates are more numerous.

3. *Astrogonium inæquale.* Pentagonal; sides rather concave. Arms acute. Dorsal ossicula rather convex, covered with small roundish granules. Marginal ossicula 2/3 on each side, the two central ones small, narrow; four others large, convex; the two at the tip very small. *Gray, P. Z. S.* 1847, p. 79. Inhab. New Guinea? or Amboina? *Capt. Sir E. Belcher.*

B. *Back rather convex; the marginal and dorsal ossicula with a small central convexity or rounded tubercle; the granules of the oral surface rather elongate, rounded.*

4. *Astrogonium tuberculatum* (T. 1. f. 2). Body pentan-

gular; sides concave; arms rather produced, acute, tapering; the ossicula of the dorsal surface, of the upper and lower marginal series, each furnished with a small, central, rounded tubercle. Marginal ossicula 2/3 on each side; the dorsal tubercles on the middle of the back and down the centre of each arm rather larger. *Gray, P. Z. S.* 1847, p. 79. Inhab. Port Natal.

C. *Body flat; ossicula of the dorsal, marginal, and oral surface entirely covered with rather elongated uniform granules; marginal ossicula small, erect, rounded above.*

5. *Astrogonium pusillosum* (T. 1. f. 1). Blackish (perhaps discoloured). Pentagonal, flat. Arm nearly as long as diameter of disk, rounded at the end. All the ossicula of the back, edge, and oral surface covered with regular, uniform, rather long, erect granules, forming a level surface; granules of the oral surface longest. The marginal ossicula narrow, erect, rounded above. Ambulacral spines elongate. *Gray, P. Z. S.* 1847, p. 79. Inhab. Port Essington.

This species, from the length of the granules, passes towards the *Astropectines*, the elongated tubercles having much the appearance of those which are called *pacilli* in that genus.

c. *Ambulacra with three or four series of equal close larger spines near the edge: body depressed, flat; marginal ossicula large, smooth, two-rowed, with only a single series of granules on each of their edges.* Gray, l. c. 1840. *Astrogonium*, sp., Müll. & Trosch. Ast. 1812.

XVI. GONIASTER. Ossicula flat; the dorsal ossicula granulated and armed with deciduous flat-based spines; both surfaces destitute of any two-lipped slits. *Gray, Ann. N. H.* 1840, p. 280.

In the younger specimens only the middle of the back and the central dorsal lines of the rays are spine-bearing; but as the animal enlarges the other tessera on the sides become covered, and at length they are separated into groups by the grooves extending from the centre to the angles of the margin between the rays. The tubercles easily fall off in the dry specimens, leaving a smooth distinct flat scar.

1. *Goniaster cuspidatus.* Body pentangular; sides curved; arms broad, triangular, rather more than half as long as the width of the body. *Gray, Ann. N. H.* 1840, p. 280. Astrogonium tessellatum, *Müll. & Trosch. Ast.* 56. Asterias tessellata? *Blainv. Actinol.* t. 23. f. 4. Pentagonaster semilunatus cuspidatus, *Linck*, 21, t. 23. f. 37 (perfect), t. 22. f. 39 (imperfect); and *Seba*, iii. t. 6. f. 9 (perfect). Ast. tessellatus, *D. and C., Lam.* Inhab. ——.

2. *Goniaster Sebæ*, Gray, Ann. N. H. 1840, p. 280. Seba, iii. t. 8. f. 2, differs in the sides of the rays being angularly inflexed.

3. *Goniaster regularis*, Gray, Ann. N. H. 1840, p. 280; Seba, iii. t. 8. f. 4, copied for *Pentagonaster regularis*, Linck, 20, t. 13. f. 22. Body with five nearly straight sides.

XVII. PENTAGONASTER. Body formed of convex, smooth, and spineless ossicula; the ossicula of the underside with a central sunk line with a central perforation and a small pit at each end. The marginal ossicula near the tips of the rays very large and swollen. *Gray, Ann. N. H.* 1840, p. 280.

1. *Pentagonaster pulchellus* (T. 8. f. 3). Body with five deeply concave sides, with four oval convex tubercles on each side, and a small one interposed between the angles of each of them. *Gray, Ann. N. H.* 1840, p. 280. Astrogonium pulchellum, *Müll. & Trosch.* 55. Asterias Mülleri, *Agassiz, MS.* (Mus. Paris). Stephanaster elegans, *Ayres, Proc. Boston Soc. N. H.* Asterias pulchella, *Gray, Encycl. Metrop.* t. .f. . Inhab. China.

When the large apical tubercles have been injured, they become divided into small unequal ones.

2. *Pentagonaster abnormalis* (T. 8. f. 1, 2). Body pentagonal; sides deeply concave, with eight narrow oblong ossicula on each side; the apical ossicula about twice the size of the others. Inhab. ——.

3. *Pentagonaster Dübeni* (T. 3. f. 2). Body flat, 5-rayed; rays two-thirds the length of the diameter of the disk, rounded at the end; ossicula all convex, rounded. Marginal ossicula ¹⁸/₉, large, round, those near the end of the arms largest and most convex. *Gray, P. Z. S.* 1847, p. 91. Inhab. Western Australia.

This species differs from *P. pulchellus* in the marginal ossicula being more equal, and in the arms being much longer and more slender. The ossicula of the dorsal disk are unequal in size and rather irregularly formed; those near the margin on the middle of the sides are oblong and narrow; those of the oral surface are more regular and not so convex, those near the angles of the mouth being the largest and subtriangular. I have named this beautiful species in memory (I regret to say) of M. W. von Düben, who recently published a very admirable paper on the northern species of this family.

See a. Astrogonium crassissimum, *Möbius, Abhandl.* iv. 1860, p. 81, t. 2. f. 1, 2, is the same or an allied species. b. Astrogonium magnificum, *Müll. & Trosch. Ast.* 53, t. 4. f. 1. c. Astrogonium astrologorum, *Müll. & Trosch. Ast.*54. d. Astrogonium geometricum, *Müll. & Trosch. Ast.* 54.

XVIII. TOSIA. The body formed of smooth and spineless ossicula, rather convex; the dorsal and ventral ossicula entire, without any impressed line, subequal; the marginal ossicula two-rowed, with a small intermediate one near each tip; dorsal wart triangular. *Gray, Ann. N. H.* 1840, p. 281.

The granules between the ossicula are deficient in the dead and washed specimens. It has been thought that the fossil species found in the chalk belonged to this genus; but the surface of the ossicula of most of the specimens I have seen show, from the scars with which their surface is covered, that they were covered with granules; therefore they rather belong to the restricted genus *Astrogonium*. *Gray, P. Z. S.* 1847, p. 80.

* The ossicula of the oral disk are more or less entirely covered with crowded, flat-topped granules.

1. *Tosia grandis* (T. 3. f. 1). Dorsal ossicula very unequal, flat-topped. Marginal ossicula ¹⁴/₄ or ¹⁶/₆ on each side, rather convex; the ossicual of the oral surface are furnished with two or three rows of crowded granules, and those near the ambulacra are most covered. *Gray, P. Z. S.* 1847, p. 80. Inhab. Western Australia.

Linck, under the name of *P. regularis*, t. 13. f. 22, 23, copied (E. M. t. 96, and *Seba*, iii. t. 8. f. 1) a species somewhat like the above, but it has only ten marginal plates. Müller, who thought he examined Linck's specimen at Leipsic, describes it as having seven upper and five under marginal plates. Goniodiscus stella, Möbius, Abhandl. iv. 1860, p. 9, t. 3. f. 1, 2, is a very nearly allied species.

2. *Tosia aurata* (T. 16. f. 2). Golden yellow. Dorsal ossicula flat-topped, the five in the centre, between the central lines of the arms, largest and round; the marginal ossicula ¹⁸/₀ or ¹²/₃, rather convex and nearly equal (that nearest the top not being longer than the others); the ossicula of the oral disk, all except a few in the middle of each area, entirely covered with flat-topped granules. *Gray, P. Z. S.* 1847, p. 80. Inhab. Australia (Brit. Mus. three spec.).

** The ossicula of the oral surface are only edged with a single series of granules, like those of the back.

3. *Tosia tubercularis* (T. 16. f. 4). Yellow, edges reddish. The dorsal ossicula convex, subtubercular, those of the centre of the arms highest, those between the arms in the centre largest, nearly flat. The marginal ossicula ⁶/₆ or ⁸/₈ on each side, convex, subtubercular; the one near the top of the arm largest and oblong, longitudinal, convex. The ossicula of the oral surface small, each surrounded with a single series of granules. *Gray, P. Z. S.* 1847, p. 91.

Var.? or young? The ossicula of the oral surface near the edges covered with granules. Inhab. Swan River.

There is a specimen in the British Museum with six marginal ossicula very like the above, but differing from it in the dorsal ossicules only being convex and rounded; it has the same convex and large marginal plate. *Gray, P. Z. S.*1847, p. 80.

4. *Tosia rubra* (T. 16. f. 3). Red brown. Dorsal ossicula rather convex, rounded. Marginal ossicula ¹⁸/₀ on each side, rather convex, equal, that at the tip of each arm smaller, narrow; the ossicula of the oral surface flat-topped, with a single series of marginal granules. *Gray, P. Z. S.* 1847, p. 81. Inhab. Australia.

5. *Tosia australis* (T. 16. f. 1). Yellowish or reddish. Dorsal ossicula rather convex, rounded. Marginal ossicula ⁶/₆ on each side, rather convex, equal; the ossicula of the oral surface flat-topped, with a single series of marginal granules. *Gray, Ann. N. H.*1840, p. 281; *P. Z. S.* 1847, p. 81. Astrogonium australe, *Müll. & Trosch. Ast.* 55. Inhab. Western Australia, Swan River.

See a. Astrogonium ornatum, *Müll. & Trosch. Ast.* 55. b. Astrogonium Lamarckii, *Müll. & Trosch. Ast.* 56.

c 2

B. Echinasterina. *The body discoidal, many-rayed; skeleton netted with numerous elongated doubly mobile articulated spines on mammillary tubercles ; dorsal warts numerous.*

XIX. Echinaster. Body star-like, granulated, depressed; back rather convex, with a circle of 10–15 conical dorsal warts! Ambulacral spines small, placed in groups with a single continuous row of large slender spines near them. The spines are very long and covered with a granular skin, and have generally a second articulation about one-third the length from the base. *Gray, Syn. Brit. Mus.* 62: *Ann. N. H.* 1840, p. 281. E. sp., *Müll. & Trosch. Ast.* 25.

1. *Echinaster Ellisii.* Dorsal warts 15; rays 11 or 12; spines large, thick. *Gray, Ann. N. H.* 1840, p. 281. Asterias echinus, *Solander and Ellis,* t. 60, 61, 62. Echinaster solaris, *Müll. & Trosch. Ast.* 25. Asterias echinites, *Lam.* Inhab. South America, *H. Cuming, Esq.*

2. *Echinaster solaris.* Rays 21; spines small; dorsal warts 10. *Gray, Ann. N. H.* 1840, p. 281. Asterias solaris, *Naturforscher,* xxviii. t. 1, 2. Inhab. ——.

C. Cribellina. *The body divided into cylindrical, elongated rays ; dorsal wart single.*

a. *Ambulacra with a single series of crowded filiform spines, sometimes united by a membrane at their base.*

† *Smooth, the rays netted, with mobile spines, with impressed dots between the network ; dorsal wart convex, flat-topped, with a few radiating grooves.*

* *Spine single, large, on the junction of the ossicula, which are placed in equidistant series.*

XX. Othilia. Skin smooth, polished; ambulacra with two very close series of filiform spines. *Gray, Ann. N. H.* 1840, p. 281. Echinaster, sp., *Müll. & Trosch. Ast.* 22.

1. *Othilia spinosa.* Rays rather more than twice the length of the width of the body. *Gray, Ann. N. H.* 1840, p. 281. Asterias spinosa, *Retz.* Pentadactylosaster spinosus, *Linck,* t. 4. f. 17. Asterias echinophora, *Lam.* n. 25, not *Chinje.* Stellaria spinosa, *Nardo, Agassiz.* Echinaster spinosus, *Müll. & Trosch. Ast.* 72. Inhab. North America, Virginia.

2. *Othilia aculeata.* Rays cylindrical, more than three or four times as long as the breadth of the body, with seven rows of acute spines. Young (or var.) arms with only five series of similar spines. *Gray, Ann. N. H.* 1840, p. 282. Inhab. Guacomayo, Central America, fine sand, 13 fathoms, *H. Cuming, Esq.*

3. *Othilia multispina.* Rays short, depressed, broad, rather more than twice as long as the width of the body, blunt at the end, with eleven rows of acute distant spines. *Gray, Ann. N. H.* 1840, p. 282 ; *Seba,* t. 7. f. 4. Echinaster brasiliensis, *Müll. & Trosch. Ast.* 72. Inhab. Brazils.

4. *Othilia purpurea.* Purplish; rays cylindrical, nearly three times as long as the width of the body, with numerous short, rather blunt spines; underside with cross wrinkles and two or three series of pores parallel to the ambulacra. Monstrosity 4-rayed. *Gray, Ann. N. H.* 1840, p. 282. Echinaster fallax, *Müll. & Trosch. Ast.* 23 ; *Savig. Descr. Egypt, Echin.* t. 4. f. 3. Inhab. "Isle of France," *W. E. Leach, M.D.*

5. *Othilia Luzonica.* Reddish brown ; rays 5 or 6, elongate, four times as long as the width of the body, with many blunt spines. *Gray, Ann. N. H.* 1840, p. 282. Inhab. Isle of Luzon, *H. Cuming, Esq.*

See a. Echinaster crassus, *Müll. & Trosch. Ast.* 23 (Mus. Paris). b. Echinaster gracilis, *Müll. & Trosch. Ast.* 23 (Mus. Paris). c. Echinaster cradanella, *Müll. & Trosch. Ast.* 24 ; New Zealand. d. Echinaster serpentarius, *Müll. & Trosch. Ast.* 24 ; Vera Cruz.

XXI. Metrodira. Slightly granular; rays slender, with large single pores and small scattered spines on the back ; smooth, and formed of regular flat ossicula on the sides. *Gray, Ann. N. H.* 1840, p. 282.

1. *Metrodira subulata.* Yellow brown; rays elongated, slender, tapering. *Gray, Ann. N. H.* 1840, p. 282. Scytaster subulatus, *Müll. & Trosch. Ast.* 36. Inhab. Mignpou, *H. Cuming, Esq.*

** *Spines small, crowded, scattered on the sides and at the junctions of the slender ossicula.*

XXII. Rhopia. Ambulacral spines long, with several series of larger spines near them. *Gray, Ann. N. H.* 1840, p. 282. Stellonia (part.), *Agassiz.*

1. *Rhopia seposita,* *Gray, Ann. N. H.* 1840, p. 282. Echinaster sepositus, *Müll. & Trosch. Ast.* 23, 126. Asterias sanguinolenta and A. sagena, *Retz. Diss.* 21, 22. Echinaster sanguinolentus, *Müll. & Trosch. Ast.* 127. Asterias seposita, *Retz. Nov. Ac.* 1783, p. 229 ; *Gmel.* 3182; *Lam.* n. 30; *Seba,* iii. t. 7. f. 5. Pentadactylosaster reticulatus, *Linck,* t. 4. f. 5. *Stellonia seposita, Nardo, Agassiz.*

2. *Rhopia Mediterranea.* Yellow ; rays 6, tapering, nearly three times as long as the width of the body ; spines short, cylindrical.

Var.? Rays 7, unequal ; spines shorter. *Gray, Ann. N. H.* 1840, p. 282. Inhab. Marseilles.

†† *Granulated, the rays above largely tubercular, not spinous, with minute dots between the tubercles, beneath uniform ; dorsal wart triangular, irregularly punctate and contorted.*

XXIII. Ferdina. Body flat; rays broad, convex and warty above, flat and uniform beneath ; ambulacral spines short, united at the base. *Gray, Ann. N. H.* 1840, p. 282.

1. *Ferdina flavescens.* Yellow, brown varied ; rays nearly half as long again as the width of the body, uniformly tubercular, blunt. *Gray, Ann. N. H.* 1840, p. 282. Inhab. Isle of France, *W. E. Leach, M. D.*

2. *Ferdina Cumingii.* Yellow or brown; rays rather longer than the width of the body, with a central and a marginal row of larger rounded tubercles and some scattered smaller ones; the larger tubercles on the sides are red when the granules are rubbed off, which they often are. *Gray, Ann. N. H.* 1840, p. 283. Inhab. West coast of Columbia, *II. Cuming, Esq.*

b. *The ambulacra with a series of very small short filiform spines (placed in pairs) with a parallel series of spines near them; the rays formed of longitudinal series of tubercles united by transverse ossicula; dorsal wart intricate.*

* *Spines near the ambulacra larger than the ambulacral ones.*

XXIV. DACTYLOSASTER. Rays cylindrical, nearly smooth, formed of regular oblong ossicula, each furnished with a central group of unequal short mobile tubercles; dorsal wart 1. *Gray, Ann. N. H.* 1840, p. 283; *Müll. & Trosch. Ast.* 33.

1. *Dactylosaster cylindricus.* Reddish, brown marbled; rays elongated, cylindrical, blunt, with eight rows of groups of spinous tubercles, three times as long as the width of the body. *Gray, Ann. N. H.* 1840, p. 283. Asterias cylindrica, *Lam.*; *Gray, Ency. Metrop.* t. .f. ; *Müll. & Trosch. Ast.* 33. Inhab. " Isle of France," *II. E. Leach, M.D.*

2. *Dactylosaster gracilis.* Reddish, brown marbled; rays slender, four times as long as the width of the body, with seven rows of groups of small spines. *Gray, Ann. N. H.* 1840, p. 283; *Müll. & Trosch. Ast.* 33. Inhab. West coast of Columbia, *II. Cuming, Esq.*

XXV. TAMARIA. Rays cylindrical, formed of seven series of granular convex roundish ossicula, each of the upper ones with three or four unequal and the lower ones with a central short blunt spine. *Gray, Ann. N. H.* 1840, p. 283; *Müll. & Trosch. Ast.* 33.

1. *Tamaria fusca.* Brown; rays rather tapering. *Gray, Ann. N. H.* 1840, p. 283; *Müll. & Trosch. Ast.* 33. Inhab. Mignpou, *II. Cuming, Esq.*

XXVI. CISTINA. Rays cylindrical, nearly smooth, formed of rows of three-lobed flat ossicula, each furnished with a central mobile spine; dorsal warts (one or two) oblong. *Gray, Ann. N. H.* 1840, p. 283; *Müll. & Trosch. Ast.* 34.

1. *Cistina Columbiæ.* Yellow; arms rather more than four times as long as the width of the body, with seven rows of spines. *Gray, Ann.N. H.* 1840, p. 283; *Müll. & Trosch. Ast.* 34. Inhab. West coast of Columbia, *II. Cuming, Esq.*
The larger specimen has two very distinct dorsal warts; but I can only see one very obscure one in the smaller specimen. It may be a monstrosity in the large specimen.

XXVII. OPHIDIASTER. Rays cylindrical, elongate, uniformly granular all over, without any spines; back with a

small central group of larger tubercles; dorsal wart concave, with radiating or twisting grooves. *Gray, Ann. N. H.* 1840, p. 383; *Agassiz.* Ophidiaster §*, *Müll. & Trosch. Ast* 27.

† *Rays cylindrical, blunt.*

1. *Ophidiaster aurantius.* Orange; rays with seven rows of rounded tubercles, about four times as long as the width of the body; spines near the ambulacra short, ovate, club-shaped. *Gray, Ann. N. H.* 1840, p. 284. ? Asterias ophidiana, *Lamk.* ii. 567. Ophidiaster ophidianus?, *Agassiz*; *Müll. & Trosch. Ast.* 28. Inhab. Madeira, rocks on Porto Santo Laurenço, *Rev. R. T. Lowe.*

2. *Ophidiaster Leachii.* Rays elongate (smooth?) with eight or nine irregular rows of unequal tubercles. The spines near the ambulacra club-shaped, rather dilated and more compressed at the tip. *Gray, Ann. N. H.* 1840, p. 284. Asterias cylindricus?, *Lamk.* ii. 567. Ophidiaster cylindricus, *Müll. & Trosch. Ast.* 29. Inhab. " Isle of France," *Dr. W. E. Leach.*

3. *Ophidiaster Guildingii,* Gray. Pale brown (dry); rays cylindrical, four times as long as the width of the body, with seven series of moderate tubercles; the spines near the ambulacra compressed, thin, ovate.
Var. 1. female? Rays thick; spaces between the tubercles large, with numerous dots.
Var. 2. male? Rays thin; spaces between the tubercles small, with four or six dots. *Gray, Ann. N. H.* 1840, p. 284.
See *a.* Oph. Hemprichii, *Müll. & Trosch. Ast.* 29; Red Sea.

†† *Rays round, tapering, acute.* Hacelia.

4. *Ophidiaster attenuatus.* Rays rounded, elongate, nearly four times as long as the width of the depressed body, broad at the base and tapering, with nine rows of triangular tubercles; spines near the ambulacra large, ovate, blunt. *Gray, Ann. H.* 1840, p. 284, *Müll. & Trosch. Ast.* 29; A. coriacea, *Grube,* 22. Inhab. ——. (Brit. Mus.)

††† *Rays triangular, tapering, with three interrupted bands of pores on each side.* Pharia.

5. *Ophidiaster pyramidatus.* Rays subangular, elongate, nearly four times as long as the width of the pyramidical body, with seven rows of tubercles; the central dorsal series much the largest; spines near the ambulacra ovate, subacute. *Gray, Ann. N. H.* 1840, p. 284; *Müll. & Trosch. Ast.* 33. Inhab. Bay of Caraccas, West Columbia, on the rocks, *II. Cuming, Esq.*

** *Series of spines near the ambulacra nearly of the same size as the aubulacral ones.*

XXVIII. LINCKIA. *Gray, Ann. N. H.* 1840, p. 284 (not Micheli). Linckia, *Nardo & Agassiz*, not Persoon nor Cuv. Ophidiaster, *Müll. & Trosch. Ast.* 30, 1843.

† *Rays 5, cylindrical, with the groups of pores scattered on the whole surface.*

1. *Linckia typus.* Pale yellow (dry); rays cylindrical.

14

elongate, rather tapering at the end, nearly seven times as long as the width of the body; back and sides with equal-sized tubercles, and moderate-sized dotted interspaces on the sides; apical tubercles moderate. *Nardo; Gray, Ann. N. H.* 1840, p. 284. *Ophidiaster miliaris, Müll. & Trosch. Ast.* 30. Pentadactylosaster miliaris, *Linck,* t. 28. f. 47. Ast. lævigata, *Linn., Lam.* 39. Distorted. Asterias cometa, *Blainville.* Inhab. Mediterranean, *Linn.* Egypt, *Sir J. G. Wilkinson.*

See *a.* Linckia franciscus, *Nardo.*

2. *Linckia crassa.* Rays elongate, thick, cylindrical, blunt at the ends, nearly three times as long as the width of the body; apical tubercle indistinct. *Gray, Ann. N. H.* 1840, p. 285. Inhab. ——?

3. *Linckia Brownii.* Rays elongate, cylindrical, rather tapering at the end, four times as long as the width of the body; back of the arms with three or four rows of small tubercles; sides with four rows of large pierced spots; apical tubercle moderate. *Gray, Ann. N. H.* 1840, p. 285; *Rumph. Amb.* t. 13. f. I? *Seba, Mus.* iii. t. 6. f. 13, 14; *Grew, Mus.* t. 8. f. 1, 2. Inhab. New Holland, *Rob. Brown, Esq.*

4. *Linckia Leachii.* Rays elongate, slender, cylindrical, rather tapering; sides with three or four rows of rather convex tubercles; apical tubercle indistinct. *Gray, Ann. N. H.* 1840, p. 285. Inhab. "Isle of France," *Dr. W. E. Leach.*

Very like *L. typus.* Our specimens, which are almost all young of the *Comet variety,* are only to be distinguished from that species by the arms being slenderer. The adult may differ more.

5. *Linckia Guildingii.* Brown, olive varied; rays slender, elongate, cylindrical, nearly equal, largely granular; back and sides with groups of three or four holes between the interspaces of the tubercles; apical tubercles large and convex. Monstrosity 6-rayed. *Gray, Ann. N. H.* 1840, p. 285; *Müll. & Trosch. Ast.* 33. Inhab. St. Vincent's, *Rev. L. Guilding.*

Differs from *L. typus* principally in being much smaller and slenderer.

6. *Linckia pacifica.* Rays elongate, cylindrical, rather tapering at the end, six times as long as the width of the body, with close oblong convex ossicula; apical tubercle indistinct; the series of spines near the ambulacra crowded together with them. *Gray, Ann. N. H.* 1840, p. 285. Inhab. Tahiti, on the reefs, *H. Cuming, Esq.*

7. *Linckia Columbiæ.* Rays elongate, cylindrical, rather tapering at the end, covered with large coarse granulations; series of spines very close to the ambulacral spines, oblong and truncated. Monstrosity with one of the rays long, the rest small, reproduced. *Gray, Ann. N. H.* 1840, p. 285. Inhab. West coast of Columbia, *H. Cuming, Esq.*

†† *Rays 5, rather trigonal, with one or two continued bands of pores, without any intervening tubercles on each side.* Phataria.

8. *Linckia unifascialis.* Rays trigonal, tapering; back

with three rows of flat ossicula; sides with a single broad band of pores; rather more than three times as long as broad. *Gray, Ann. N. H.* 1840, p. 285. Ophidiaster suturalis?, *Müll. & Trosch. Ast.* 30. Inhab. Bay of Caracens, West Columbia, on the rocks at low water, *H. Cuming, Esq.*

9. *Linckia bifascialis.* Rays trigonal; back with four or five rows of irregular convex ossicula at the base, and many at the end of the ray; sides of the ray with two broad bands of pores at the base and one at the end. *Gray, Ann. N. H.* 1840, p. 285.

††† *Rays depressed, with a single pore between each dorsal ossiculum, and a narrow band of a few pores along each side of the arm.* Acalia.

10. *Linckia pulchella.* Brown; rays flat, nearly three times as long as the width of the body; the spines near the ambulacra oblong, compressed, truncated. *Gray, Ann. N. H.* 1840, p. 226; *Müll. & Trosch. Ast.* 37. Inhab. ——.

11. *Linckia intermedia.* Rays elongate, cylindrical, rather tapering at the end, formed of oblong convex ossicula; pore on the back single, on the sides in two rows of groups of three or four; the series of spines on the side of the ambulacra separate from it and from one another. *Gray, Ann. N. H.* 1840, p. 286; *Müll. & Trosch. Ast.* 37. Inhab. ——.

12. *Linckia erythræa.* Rays elongate, cylindrical; the row of small spines near the ambulacra double in some part of its length. *Gray, Ann. N. H.* 1840, p. 286; *Müll. & Trosch. Ast.* 37. Inhab. Red Sea, *James Burton, Esq.*

See *a.* Ophidiaster diplax, *Müll. & Trosch. Ast.* (Mus. Berlin). *b.* Ophid. ornithopus, *Müll. & Trosch. Ast.* 31; Vera Cruz. *c.* Asterias multiforas, *Lamk.* ii. 565; Ophid. multiforus, *Müll. & Trosch. Ast.* 31. *d.* Ophid. Ehrenbergii, *Müll. & Trosch. Ast.* 31 (Berl. Mus.). *e.* Ophid. tuberculatus, *Müll. & Trosch. Ast.* 32 (Berl. Mus.). *f.* Ophid. echinulatus, *Müll. & Trosch. Ast.* 32 (Mus. Leyden).

e. Ambulacra with a series of short filamentous spines, placed in groups of four or five (one group on each ossiculum); rays formed of series of tubercles, with (one or two) small holes between them, and covered with granules.

* *Rays with only one (or two) series of small spines on the side of the ambulacral spines.*

XXIX. Fromia. Rays 5–8, flat, triangular, formed of flat-topped granular tubercles. *Gray, Ann. N. H.* 1840, p. 286.

1. *Fromia milleporella,* Gray, Ann. N. H. 1840, p. 286. *Scytaster posterius, Müll. & Trosch. Ast.* 35. *Asterias Sebæ,* Blainv., Seba, Thesaur. t. 8. f. a, b. Var. 1. Rays 6, rather slender. Var. 2. Rays 7, slenderer. Var. 3. Larger, 5- or 6-rayed. Inhab. Isle of

France, *Dr. W. E. Leach*; Indian Ocean, *Gen. Hard-wicke*; Red Sea, *James Burton, Esq.*
See *a.* Scytaster semiregularis, *Müll. & Trosch. Ast.* 36.
b. Scytaster Kcenii, *Müll. & Trosch. Ast.* 38.

XXX. GOMOPHIA. Rays elongate, cylindrical, tapering, with a terminal tubercle; back with large rounded tubercles; back of the rays with series of large conical convex tubercular spines; the spines near the ambulacra small, crowded. *Gray, Ann. N. H.* 1840, p. 286.

1. *Gomophia Egyptiaca.* Rays tapering, acute, four times as long as the width of the body, with five irregular rows of conical acute tubercles. *Gray, Ann. N. H.* 1840, p. 286. Inhab. Egypt, *Sir J. G. Wilkinson.*

** *Rays with the series of spines on the sides of the ambulacra gradually passing into the granulations which crowd on them.*

XXXI. NARDOA. Rays cylindrical, spineless, formed of large granular convex ossicula. *Gray, Ann. N. H.* 1840, p. 286. Scytaster, sp., *Müll. & Trosch. Ast.* 31.

1. *Nardoa variolata,* Gray, Ann. N. H. 1840, p. 286. Scytaster variolatus, *Müll. & Trosch. Ast.* 31. Asterias variolatus, Retz. Diss. 19; Lamk. 36; Oudart, t. . f. . Pentadactylosaster variolatus, Linck, t. 1. f. 1, t. 8. f. 10, t. 11. f. 24. Linckia variolata, Nardo. Inhab. Mediterranean Sea.

2. *Nardoa Agassizii.* Rays cylindrical; tubercles subequal.
Var. 1. 4-rayed, *Linck,* t. 1. f. 1. Var. 2. 6-rayed. Monstrosity 1. 7-rayed. Monstrosity 2. 3-rayed, with two short rays on the opposite side. Monstrosity 3. with one ray bifid, *Linck,* t. 14. f. 2, 4. *Gray, Ann. N. H.* 1840, p. 287. Inhab. Isle of France, *Dr. W. E. Leach.*

3. *Nardoa tuberculata.* Rays cylindrical, with scattered hemispherical larger tubercles. *Gray, Ann. N. H.* 1840, p. 287. Inhab. Island of Luzon, Port of Sual, *H. Cuming, Esq.*
See *a.* Asterias milleporella, *Lamk.* ii. 564; Scytaster milleporellus, *Müll. & Trosch. Ast.* 35; Red Sea. *b.* Scytaster zodiacalis, *Müll. & Trosch. Ast.* 35.

XXXII. NARCISSIA. Body pyramidical, thin, coriaceous, uniformly granular; rays tapering, elongate, triangular on the base, formed of thin flattened ossicula. *Gray, Ann. N. H.* 1840, p. 287.

1. *Narcissia Teneriffæ.* Rays tapering, elongate, acute, more than four times as long as the width of the body. Inhab. Teneriffe (Brit. Mus.).

XXXIII. NECTRIA. Body rather pyramidical, coriaceous, scattered with truncated warts, granular at the top; rays roundish, produced, edged with two series of flat granular warts on each side; beneath largely granular. *Gray, Ann. N. H.* 1840, p. 287.

1. *Nectria ocellifera,* Gray, Ann. N. H. 1840, p. 287. Asterias oculifera, Lamk. n. 5; Oudart, t. . f. . Goniodiscus ocelliferus?, Müll. & Trosch. Ast. 60. Inhab. ——. (Brit. Mus.).

XXXIV. NEPANTHIA. Body small, flat; rays very long, cylindrical, tapering, not margined, formed, above and below, of many regular longitudinal and transverse series of flat-topped tubercles, furnished at the top with a series of elongate spine-like granulations. *Gray, Ann. N. H.* 1840, p. 287.
Intermediate between *Astropectinidæ* and *Cribellinæ*; but the rays are not margined, and the spines at the top of the tubercles are not regularly radiately disposed.

1. *Nepanthia tessellata.* Brown; rays elongate, slender, tapering, with series of square warts. *Gray, Ann. N. H.* 1840, p. 287; *Müll. & Trosch. Ast.* 28. Inhab. ——. (Brit. Mus.).

2. *Nepanthia maculata.* Grey, with black spots; rays rather depressed, blunt; middle of the back with oblong transverse, and the sides with squarish, warts. *Gray, Ann. N. H.* 1840, p. 287; *Müll. & Trosch. Ast.* 28. Inhab. Migupon, *H. Cuming, Esq.*

Family IV. ASTERINIDÆ.

Body discoidal or pyramidical, sharp-edged; skeleton formed of flattish imbricate plates; dorsal wart single, rarely double. *Gray, Syn. Mus.* 62; *Ann. N. H.* 1840, p. 288. Asteriscus, *Müll. & Trosch. Ast.* 39.

A. *Ambulacral spines separate.*

I. PALMIPES. Body flat, thin, nearly membranaceous; margin radiately striated; the dorsal ossicula with a radiating tuft, and the oral ones with a transverse line of many thin mobile spines; ambulacral spines in oblique rounded groups. *Gray, Ann. N. H.* 1840, p. 288; *Agassiz, Prod.* 25.

1. *Palmipes membranaceus.* Rays 5, broad. *Gray, Ann. N. H.* 1840, p. 288; *Linck,* t. 1. f. 2, 3. Asteriscus palmipes, *Müll. & Trosch. Ast.* 39. Ast. membranacea, *Retz. & Lamk.* Ast. placenta, *Pennant.* Ast. cartilaginea, *Fleming.* Ast. rosacea, *Lamk.* (a broken specimen). Inhab. British Channel, Plymouth Sound; Mediterranean?

2. *Palmipes Stokesii.* Rays 15, acute. *Gray, Ann. N. H.* 1840, p. 288. Asterias rosacea, var., *Lamk.* ii. 558. Asteriscus rosaceus, *Müll. & Trosch. Ast.* 40 (Mus. Mr. Stokes). Inhab. Japan.
See *a.* Asteriscus pectinifer, *Müll. & Trosch. Ast.* 40.

II. PORANIA. Body pyramidical, thick, 5-rayed; skin above and below varnished, spineless; dorsal ossicula irregular; the margin with two series of large ossicula; the lower ones produced, sharp-edged, and each furnished on the edge with a series of mobile spines; the ambulacra with two series of mobile spines, each pair on a separate ossiculum; the upper marginal ossicula trigonal, imbricate;

16

the dorsal ones unequal, irregular; the central of the lower marginal ossicula with four and the apical ones with a pair of spines. Allied to *Gymnasteria.* Gray, Ann. N. H. 1840, p. 288. *Asteropsis*, sp., Müller.

1. *Porania gibbosa*, Gray, Ann. N. H. 1840, p. 288. *Asterias pulvillus*, O. Müller. Zool. Dan. t. 19. *Asteropsis pulvillus*, Müll. & Trosch. Ast. 64, t. 8. f. 2, 128. *Asterias gibbosus*, Leach. Brit. Mus. 1817. *Ast. equestris*?, Thompson, Mag. Nat. Hist.iv. 237. *Goniaster Templetoni*, Forbes, Wern. Trans. 1839, p. 6, t. 4. f. 1, 2. Inhab. Isle of Arran and Plymouth Sound, *Dr. W. E. Leach*, 1817. Isle of Man, Douglas Bay, *J. R. Wallace, Esq.*

See *a.* Asterias vernicina, *Lamk. Hist.* ii. 554. Asteropsis vernicina, *Müll. & Trosch. Ast.* 66; Panama.

III. ASTERINA. Body rather pyramidical, 5-rayed; the back convex; the oral surface flat; the ossicula of each surface furnished with one or more mobile tapering spines; the margin sharp-edged; each of the ossicula with a marginal series of spines; ambulacral spines placed in groups of four or five. *Gray, Ann. N. H.* 1840, p. 289; *Nardo.*

1. *Asterina gibbosa.* Each of the ossicula of the oral surface with a central pair of mobile tapering spines. Each of the marginal ossicula of the dorsal surface with a pair of spines, of the discal one with many crowded pairs; back with series of distinct pores. *Gray, Ann. N. H.* 1840, p. 289; *Forbes.* Asterias gibbosa, *Pennant, B. Z.* iv. 121. n. 6; *Flem. B. A.* 486. Pentaceros plicatus et concavus, *Linck,* 25, t. 3. f. 20. Asteriscus exigua, *Pet. Gaz.* t. 16. f. 8. *Ast.* minuta, *Linn.?* Ast. stellata obtusa ciliata, *Linn. F. Suec.* 2112. Asterina minuta, *Agassiz?* Asterias pulchella, *Blainr.?, Faun. Franc.* t. ; *Man. Malac.* t. 22. f. 8. Asterias vermiculata, *Müll. & Trosch. Ast.* 41. Inhab. Plymouth Sound, *Dr. W. E. Leach;* Ireland, *Linck;* Marseilles, *Dr. W. E. Leach;* Sicily, *W. Swainson, Esq.;* Madeira, *Rev. — Bulwer.*

2. *Asterina Burtonii.* Rays elongate, convex, blunt at the end; each of the ossicula of the oral surface with a central group of three crowded mobile tapering spines; of the dorsal surface with a crowded group of short tubercles. *Gray, Ann. N. H.* 1840, p. 289. Inhab. Red Sea, *James Burton, Esq.*

3. *Asterina minuta.* Each of the ossicula of the oral surface with a single spine or a central group of three crowded mobile spines; of the dorsal surface granular, with a few very small spicula on the upper edge; and of the margin with a spreading tuft of spines. *Gray, Ann. N. H.* 1840, p. 289. Asteriscus minutus, *Müll. & Trosch. Ast.* 41. Asterias minuta, *Linn.; Gmelin?* Asterias exigua, *Lamk.* n. 43; *Sebu,* iii. t. 5. f. 15.

Var. 1. Larger; each of the ossicula of the oral surface with three spines. Var. 2. Smaller; each of the ossicula with one, rarely with two spines. Monstrosity 1. Rays 4; and 2, rays 6. Inhab. America, *Linn.* West Indies, St. Vincent's, *Rev. L. Guilding.*

The specimens of the two varieties exactly resemble each other except in the characters mentioned, and they appear to have been taken at the same time.

4. *Asterina Krausii.* Olive-green; the central ossicula of the oral surface spineless, those near the margin with a single central triangular spine; the dorsal ossicula with a series of minute, very short blunt spines. *Gray, Ann. N. H.* 1840, p. 289. Asteriscus Kraussi, *Müll. & Trosch. Ast.* 42; *E. M.* t. 100. f. 4, 5. Inhab. Cape of Good Hope, *Dr. Kraus.*

5. *Asterina Gunnii.* The central ossicula of the oral surface with one and the marginal ones with a pair of cylindrical blunt spines; the dorsal ossicula with radiating groups of short cylindrical spinulose spines; body with six slightly concave sides. *Gray, Ann. N. H.* 1840, p. . Asteriscus australis, *Müll. & Trosch. Ast.* 43. Var. Body five-sided. Var. or Monstrosity with two dorsal warts. Inhab. Van Diemen's Land, *Ronald Gunn, Esq.*

6. *Asterina calcar.* All the ossicula of the lower surface with a single central cylindrical blunt spine; the dorsal ones with numerous short tapering spinulose spines; body convex, with 8 rather elongate blunt rays. *Gray, Ann. N. H.* 1840, p. . Asterias calcar, *Lamk.* 17; *Oudart,* t. . f. . Inhab. Van Diemen's Land, *Dr. Lhotsky* and *Mr. G. B. Sowerby.*

See *a.* Asteriscus Cepheus, *Müll. & Trosch. Ast.* 41. *b.* Asteriscus pentagonus, *Müll. & Trosch. Ast.* 42; *Seba,* t. 5. f. 13; *E. M.* t. 100. f. 3. *c.* Asterias penicillaris, *Lamk.* ii. 555; Asteriscus penicillaris, *Müll. & Trosch. Ast.* 42. *d.* Asteriscus Diesingi, *Müll. & Trosch. Ast.* 43.

IV. PETRICIA. Body convex, 5-rayed. Skin above and below varnished and spineless. Back strengthened with numerous, sunken, moderate-sized ossicula; the margin with two series of larger oblong ossicula, but spineless; the oral surface with rather regularly disposed smaller ossicula. Ambulacral spines subulate, placed in pairs, with a second series of similar but rather larger spines on the outer side of them. *Gray, P. Z. S.* 1840, p. 80.

This genus is very like *Porania,* but the back does not appear to be angular, the margin is edged with spines, and the ambulacral spines are in pairs, and not single as in that genus. The ossicula of the back and oral surface are punctured; and one of them, situated near the edge of the back, in the middle space between the arms, is furnished with a linear pore edged with convex lips.

1. *Petricia punctata* (T. 6. f. 1). Orange when dry. *Gray, P. Z. S.* 1847, p. 80; *Ann. N. H.* 1847, p. 202. Inhab. the Reef of Attagor, *J. B. Jukes, Esq.*

There is a single specimen of this species in the British Museum collection.

V. PATIRIA. The body pyramidical, coriaceous, with 5 rays: the ossicula of the oral surface with uniform radiating groups of small spines; of the dorsal surface of two kinds—the one crescent-shaped with series of small bundles of spines, the others bearing irregular round bundles of spines between them. The upperside, between the angles of the arms, is covered with small, roundish groups of spines. *Gray, Ann. N. H.* 1840; *P. Z. S.* 1847, p. 82.

1. *Body pentagonal; the dorsal ossicula lunate, narrow; the edge of the arms acute.*

1. *Patiria coccinea.* Scarlet, the body 5-rayed, sides concave, the end of the rays rather slender, blunt. *Gray, Ann. & Mag. N. H.* 1810; *P. Z. S.* 1847, p. 82. *Asteriscus coccineus, Müll. & Trosch.* 43.

The roundish group of spines between the lunate ossicula are very abundant. Inhab. Cape of Good Hope.

2. *Body 5-rayed; rays thick, rounded; dorsal ossicula lunate, subtriangular; arms convex above and rounded on the sides.*

2. *Patiria granifera.* Brown. Back rather convex. The arms broad, rounded at the end, nearly as long as the diameter of the disk, rounded above, flat beneath; the lunate dorsal ossicula covered with short, crowded spines, and with only a few small tufts of spines between them; the ossicula of the oral surface each with a transverse line of six or eight spines. *Gray, P. Z. S.* 1847, p. 82. Asterias granifera?, *Lam.* n. 24?; var. à petits grains, *Oudart,* t. Inhab. ——?

Variety. The arms more slender, about one-third longer than the diameter of the disk. Inhab. ——? (Brit. Mus.).

The variety may be a distinct species; but the specimen is not in sufficiently good preservation to determine this point with certainty.

3. *The body 5-rayed, rays thick, rounded; the dorsal ossicula, especially those at the end of the arms, broad, rounded; the back covered with two or three beaked pedicellaria nearly hiding the tubercles.*

3. *Patiria ocellifera.* Body 5-rayed; arms thick, rounded, as long as the diameter of the disk, bluntish at the end; the dorsal ossicula broad, oblong or roundish, reddish, covered with short, crowded spines; the oral surface with transverse rows of three to five mobile spines. *Gray, P. Z. S.* 1847, p. 82. ?? Goniodiscus ocelliferus, *Müll. & Trosch. Ast.* 60. Asterias ocellifera, *Lam.* 45; *Oudart,* t. . fig. . Inhab. ——?

This species much more nearly resembles Oudart's figure than the species I have described under the name of *Nectria oculifera.*

4. *Patiria obtusa.* Brown, depressed, 5- to 6-rayed; rays depressed, rounded at the end; dorsal surface with lunate ossicula crowded with short spines; oral surface with circular groups of crowded spines in the middle of each ossiculum. *Gray, P. Z. S.* 1847, p. 82. Inhab. Panama. Sandy mud, 6 to 10 fathoms.

5. *Patiria? crassa.* Pale yellow (dry), 5-rayed; rays thick, rather tapering, about half as long again as the diameter of the disk. Dorsal surface formed of convex,
subhemispherical ossicula, covered with crowded minute spines. The oral surface with roundish groups of short, crowded spines, like *paxilli. Gray, P. Z. S.* 1847, p. 83. Inhab. Western Australia (*Mr. Gould*).

See *a.* Asteriscus setaceus, *Müll. & Trosch. Ast.* 43. *b* Asterias trochiscus, *Retz. Diss.* 10; Asteriscus trochiscus, *Müll. & Trosch. Ast.* 43. *c.* Asterias miliaris, *Müller, Zool. Dan.* t. 131 (not text); Asteriscus miliaris, *Müll. & Trosch. Ast.* 42; North Sea.

B. *Ambulacral spines radiating, webbed together.*

VI. PTERASTER. Ambulacra edged with a series of radiating, webbed spines. *Pteraster,* Müll. & Trosch. Ast. App. 127, 1843.

1. *Pteraster miliaris,* Müll. & Trosch. Ast. 128, t. 6. f. 1. *Asterias miliaris,* O. F. Müller, Zool. Dan. t. 131 (not text). *Asteriscus miliaris,* Müll. & Trosch. Ast. 44. Inhab North Sea.

2. *Pteraster Capensis.* Body subpentagonal, swollen, edge very thick, rounded; back convex, reticulated, with rounded groups of very small ossicula at the junction of the reticulations. *Gray, P. Z. S.* 1847, p. 82. Inhab. Cape of Good Hope.

The spines of the ambulacra are like those of *Pteraster miliaris,* but they are longer, and the series of webbed spines on their outer margins are scarcely longer than those of the ambulacra, while in the northern species they are much longer and thicker, and there is no appearance of the two long glassy spines at the angle of the mouth so distinct and peculiar in that species.

VII. GANERIA. Body flat, 5-rayed. Back coriaceous, strengthened with numerous small, linear and curved series of very short cylindrical spines. Margin perpendicular, with two series of narrow ossicula, each armed with a central, erect linear series of short cylindrical spines. Oral surface covered with diverging spines, one being placed on each ossiculum. Ambulacra linear, with two series of tentacles, and edged with subulate spines, two on each ossiculum, and with a series of diverging spines at the angles near the mouth. *Gray, P. Z. S.* 1847, p. 83.

1. *Ganeria Falklandica.* Body 5-rayed; rays as long as the diameter of the disk, rather blunt at the tip. *Gray, P. Z. S.* 1847, p. 83. Inhab. Falkland Islands (*Captain Sir James Ross*).

VIII. SOCOMIA. The body depressed; rays elongate formed of imbricate plates; the margins broad, the upper and lower series of ossicula being separated by a groove. *Gray, Ann. & Mag. N. H.* 1840.

1. *Socomia paradoxa.* Yellow. Inhab. ——? *Gray, Ann. & Mag. N. H.* 1840.

D

EXPLANATION OF PLATES.

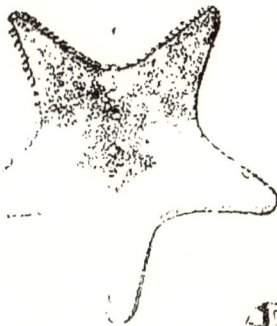

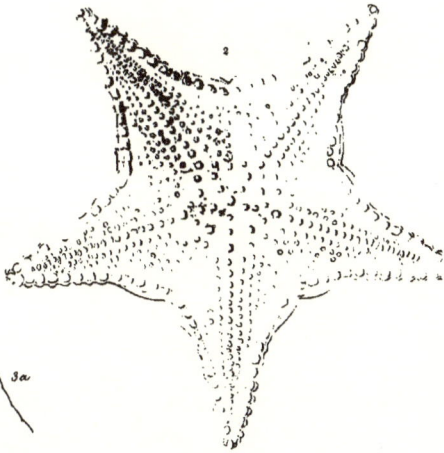

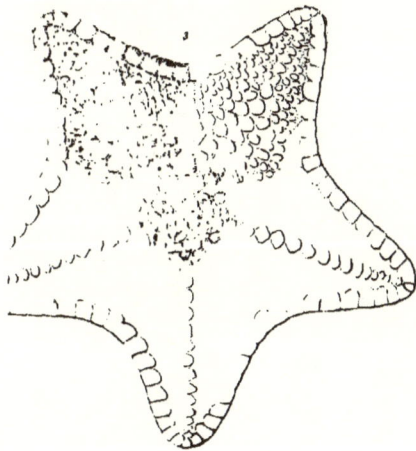

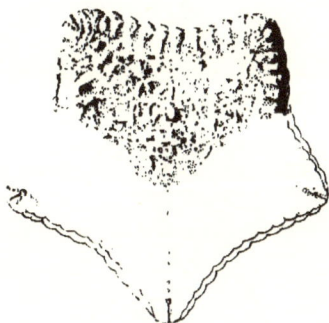

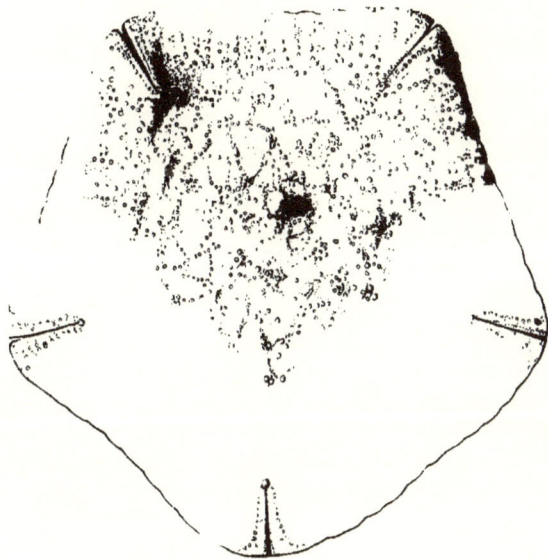

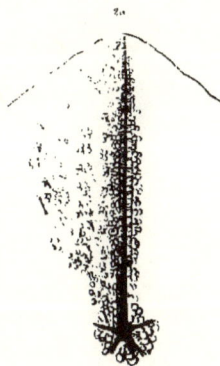

William Wing del. et lith.   Hullmandel & Walton Lithographers

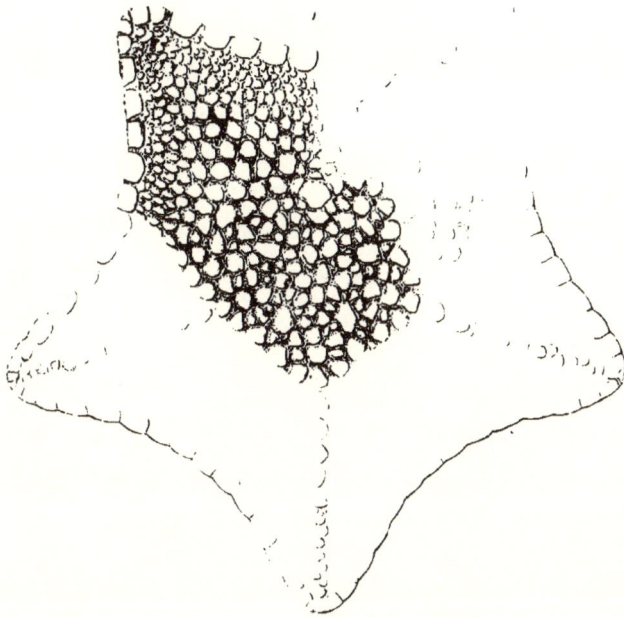

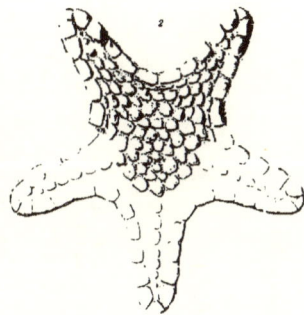

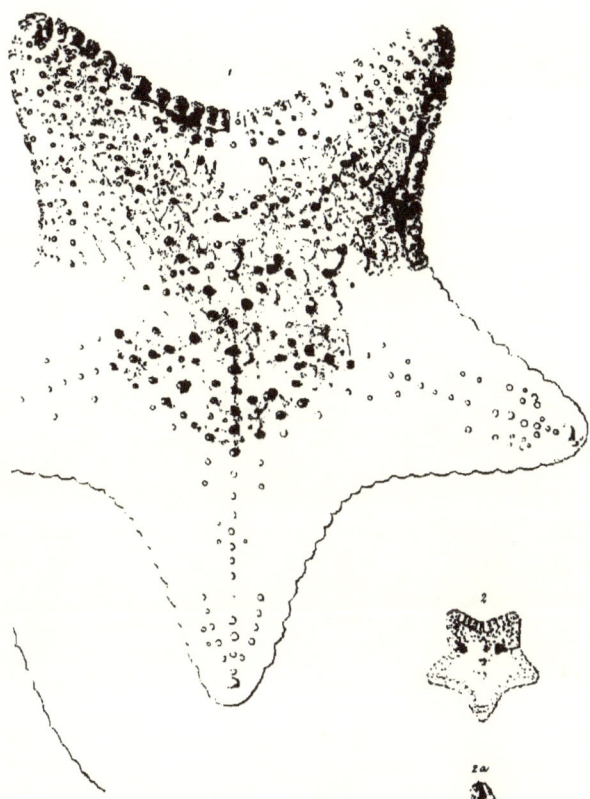

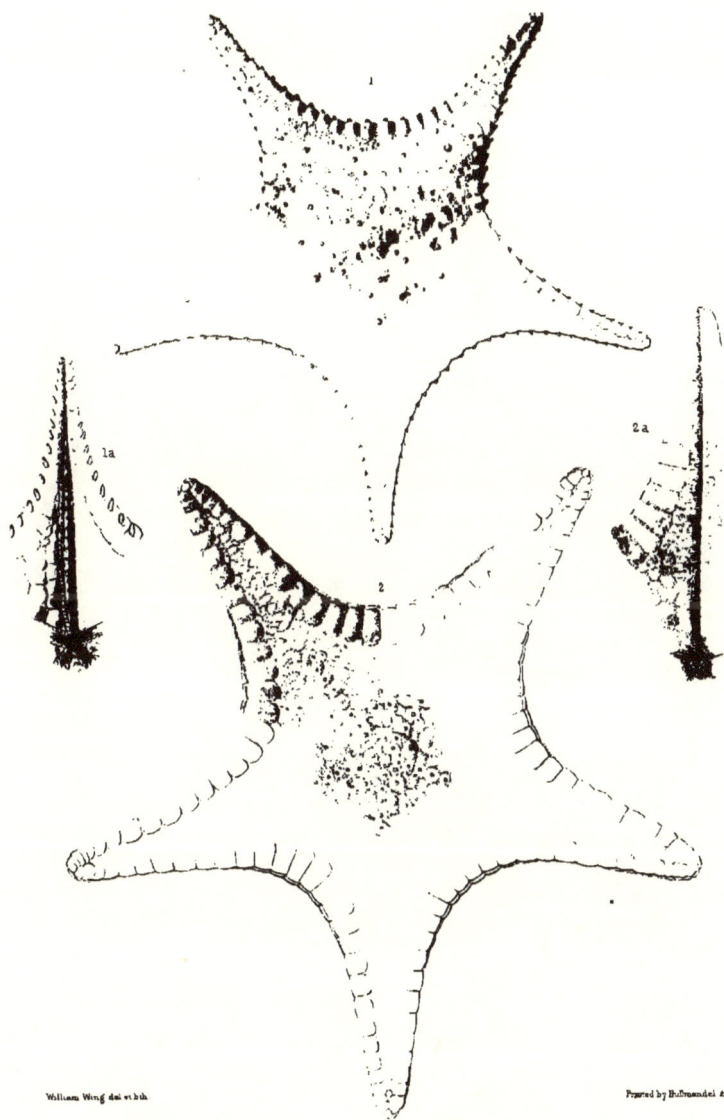

1 STELLASTER INCII   2 ANTHENEA GRANULIFERA

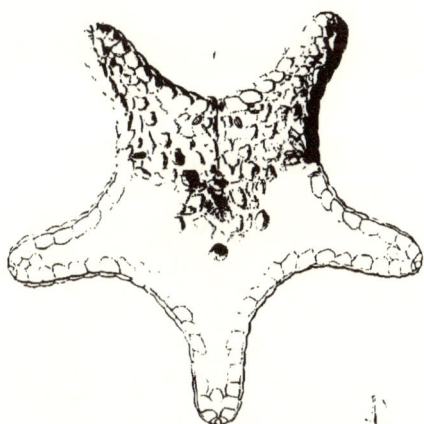

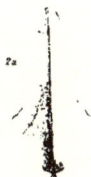

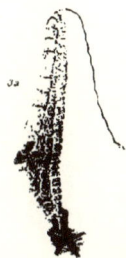

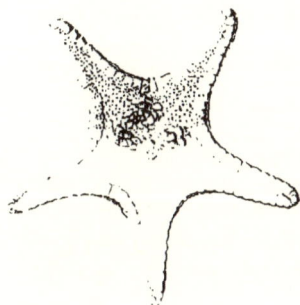

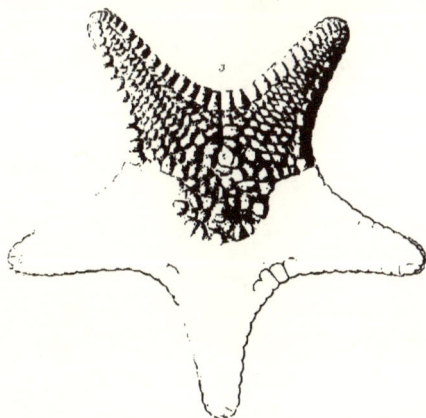

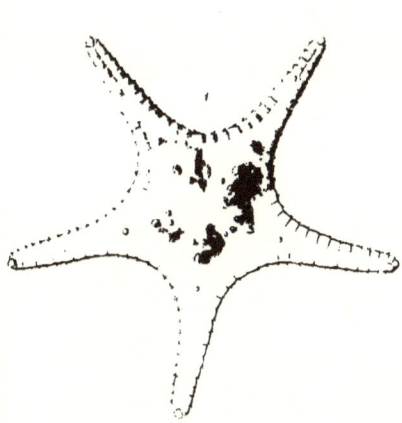

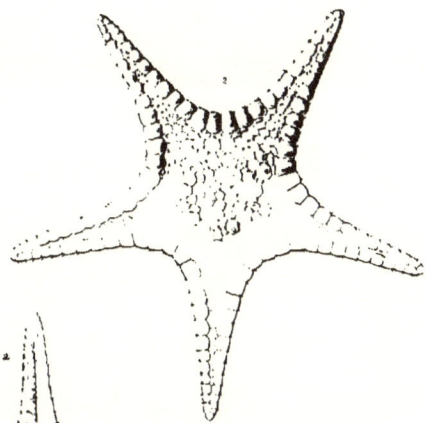

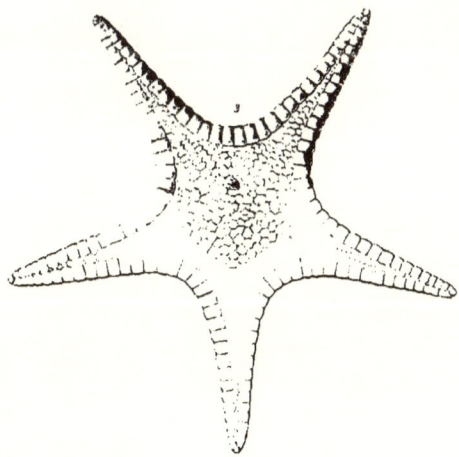

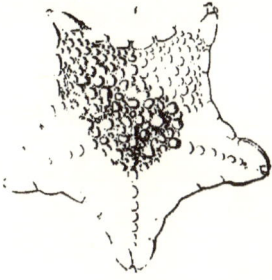

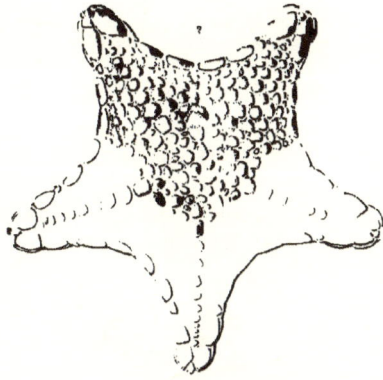

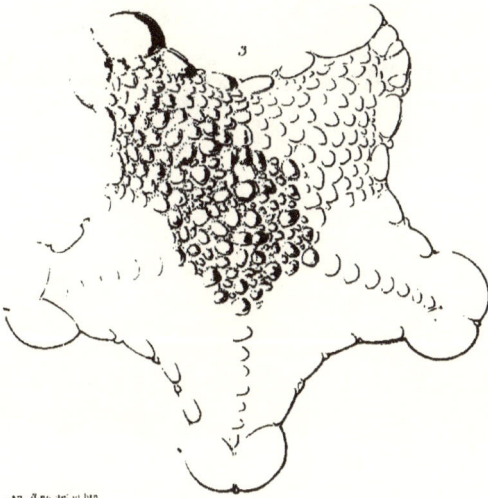

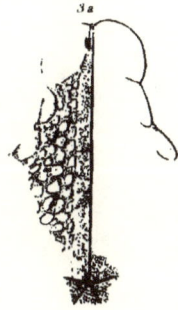

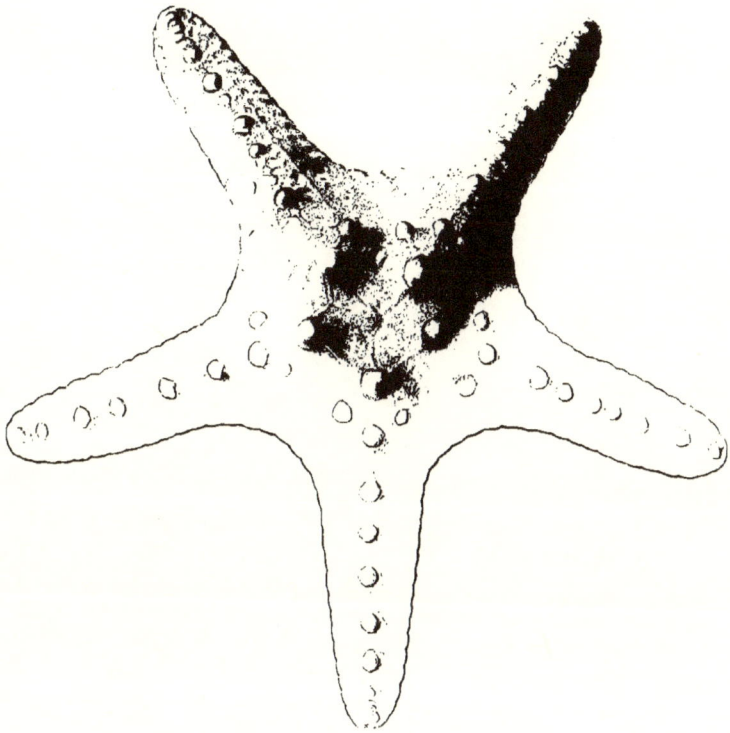

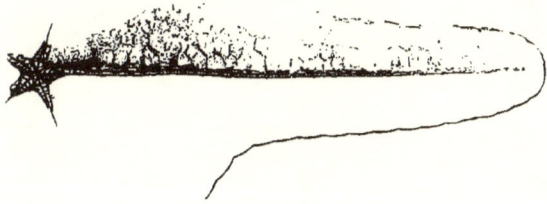

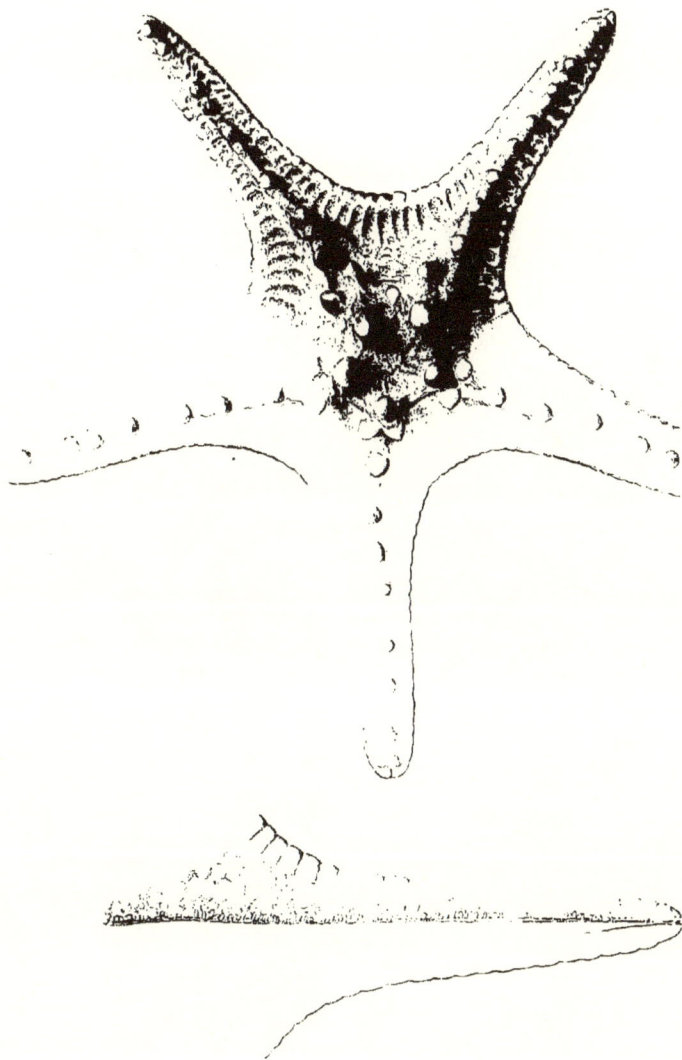

PENTACEROS.

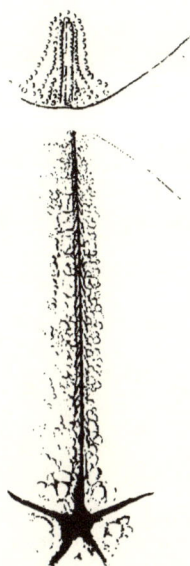

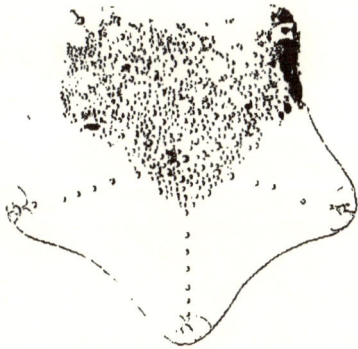

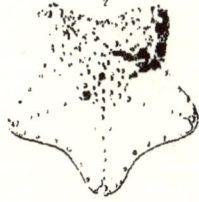

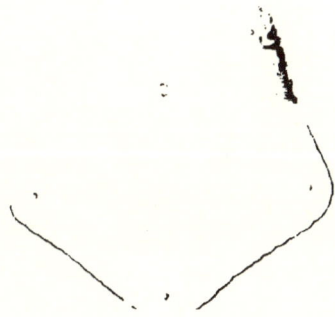

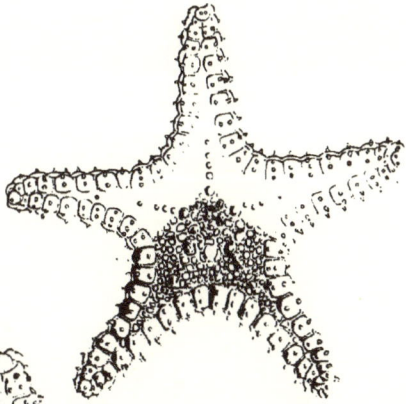

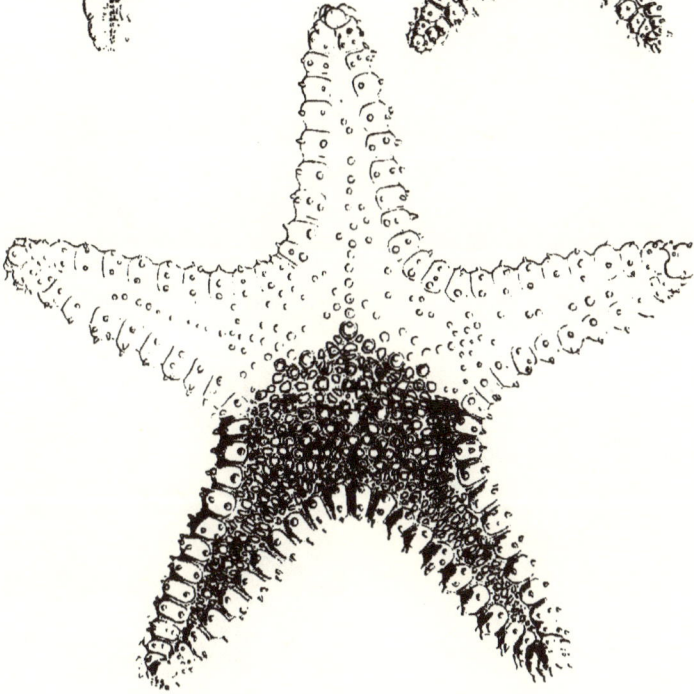

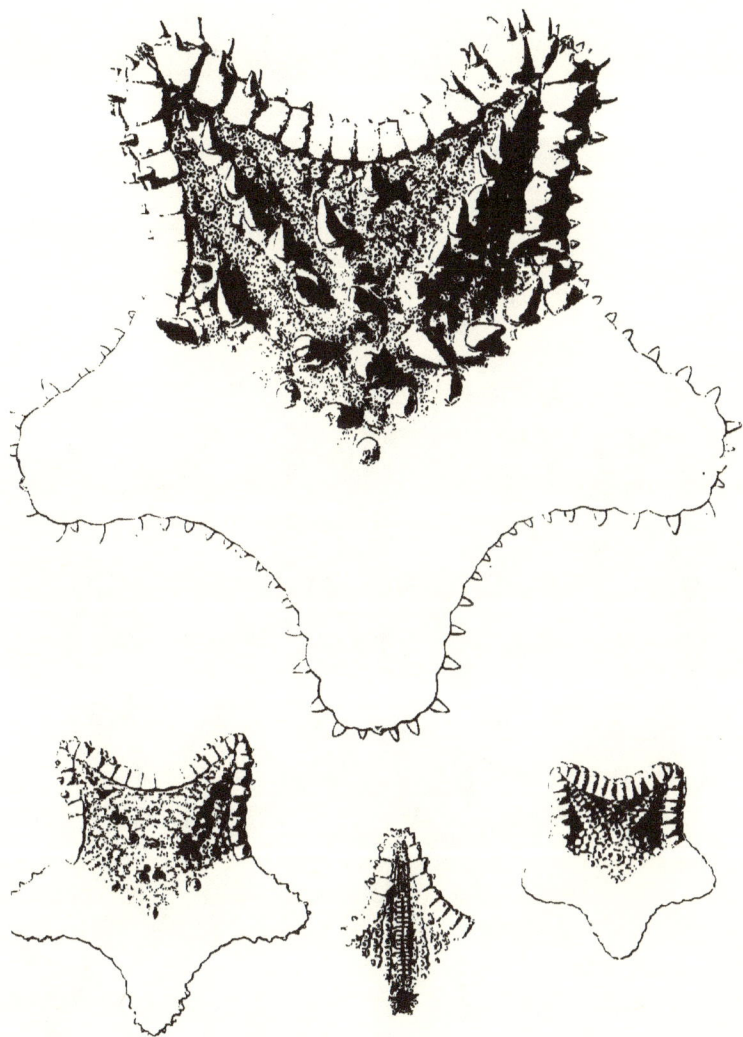

Tab. 15.

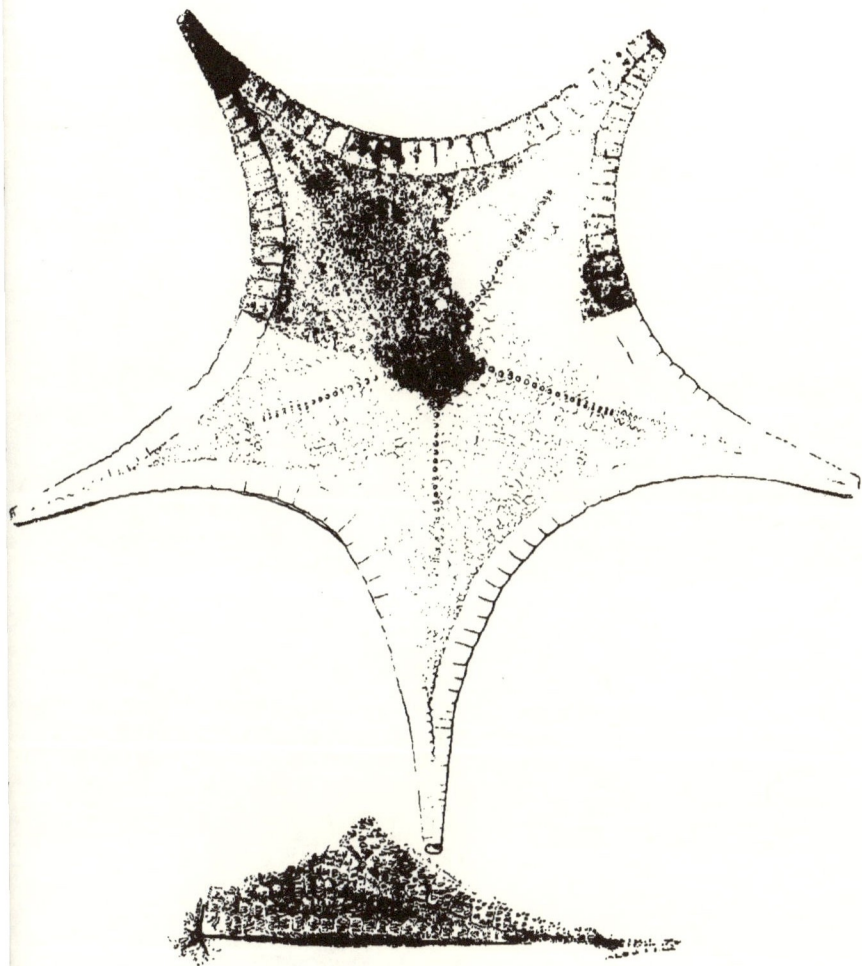

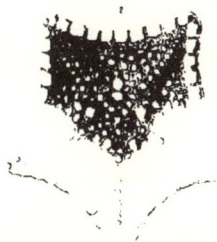